Some Aspects of Diamonds in Scientific Research and High Technology

Edited by Evgeniy Lipatov

Published in London, United Kingdom

IntechOpen

Supporting open minds since 2005

Some Aspects of Diamonds in Scientific Research and High Technology
http://dx.doi.org/10.5772/intechopen.78882
Edited by Evgeniy Lipatov

Assistant to the Editor(s): Dmitriy Genin

Contributors
Awadesh Kumar Kumar Mallik, José Vieira, Mariana Amorim Fraga, Vladimir Jesus Trava-Airoldi, Ikuko Akimoto, Nobuko Naka, Karin Larsson, Suya Prem Anand P, Evgeniy Lipatov

First published in London, United Kingdom, 2020 by IntechOpen
IntechOpen is the global imprint of INTECHOPEN LIMITED, registered in England and Wales, registration number: 11086078, 7th floor, 10 Lower Thames Street, London, EC3R 6AF, United Kingdom
Printed in Croatia

British Library Cataloguing-in-Publication Data
A catalogue record for this book is available from the British Library

Additional hard and PDF copies can be obtained from orders@intechopen.com

Some Aspects of Diamonds in Scientific Research and High Technology
Edited by Evgeniy Lipatov
p. cm.
Print ISBN 978-1-78984-516-7
Online ISBN 978-1-78984-517-4
eBook (PDF) ISBN 978-1-83962-596-1

For EU product safety concerns:
IN TECH d.o.o., Prolaz Marije Krucifikse Kozulić 3, 51000 Rijeka, Croatia, info@intechopen.com or visit our website at intechopen.com.

Meet the editor

Evgeniy Lipatov graduated from the Radiophysical Faculty of Tomsk State University in 2000. He defended his thesis for the degree of Candidate of Physical and Mathematical Sciences in the specialty "Optics" at Tomsk State University in 2012. At present he is Head of the Laboratory of Quantum Information Technologies of Tomsk State University, senior researcher at the Institute of High Current Electronics, and assistant professor at the Department of Quantum Electronics and Photonics at Tomsk State University.

Contents

Preface XI

Chapter 1 1
Introductory Chapter: Some Aspects of Diamonds in Scientific Research and High Technology
by Evgeniy Lipatov and Dmitrii Genin

Chapter 2 7
Polycrystalline Diamond Characterisations for High End Technologies
by Awadesh Kumar Mallik

Chapter 3 35
Significance of Diamond as a Cutting Tool in Ultra-Precision Machining Process
by P. Suya Prem Anand

Chapter 4 51
Development, Properties, and Applications of CVD Diamond-Based Heat Sinks
by José Vieira da Silva Neto, Mariana Amorim Fraga and Vladimir Jesus Trava-Airoldi

Chapter 5 69
Intrinsic Carrier Parameters and Optical Carrier Injection Method in High-Purity Diamonds
by Ikuko Akimoto and Nobuko Naka

Chapter 6 85
Simulation of Diamond Surface Chemistry: Reactivity and Properties
by Karin Larsson

Preface

Nowadays, many researchers all over the world consider diamond as a wide-gap semiconductor material. Its key parameters, such as charge carrier mobility, breakdown voltage, and transparency in the ultraviolet, visible, and infrared ranges, are higher than those of other semiconductors. Only gallium arsenide and graphene have larger values of electron mobility. For years it was difficult to use diamonds in electronics and photonics due to the low quality of natural and synthetic samples.

However, the technologies of chemical vapor deposition diamond synthesis and doping enable us to apply diamond-based devices in high technologies. Besides that, diamond shows good results as a material for heat sinks. Due to its extremely high hardness it has perspectives in the production of high-tech optical elements, intendors, and anvils. The spectrum of possible diamond applications is permanently widening.

This book describes some aspects of synthetic diamond applications in the high-tech field. It will be interesting and useful for master's course students, researchers, and technologists who are investigating and working with this material.

Evgeniy Lipatov
Tomsk State University,
Institute of High Current Electronics,
Tomsk, Russian Federation

Chapter 1

Introductory Chapter: Some Aspects of Diamonds in Scientific Research and High Technology

Evgeniy Lipatov and Dmitrii Genin

1. Introduction

Nowadays, the diamond is the most promising and unused semiconductor material in electronics and photonics due to the technological limitations of large sample production [1]. However, diamond is used for drilling, grinding, and polishing hard rocks and structural materials. Diamond itself can be mechanically processed using the difference of its wear resistance for different crystallographic directions.

In addition to the diamond mining industry, the volume of synthesized rough diamonds is steadily increasing. Alrosa Group of Companies (Yakutia, Russia) is the world largest natural diamond supplier. It mines about 35–40 million carats (7–8 tons) per year [2], which makes up about 28% of world production. At the same time, diamond plants located in China annually produce ~10 billion carats (~2000 tons) of synthetic diamonds, which are mainly used as abrasive material [3].

Natural diamond is a mineral that was formed at high pressures and high temperatures (HPHT) and annealed under HPHT conditions for a long time. Usually, natural diamond contains a large amount of nitrogen in the form of polyatomic impurity-defective centers [4]. In addition, the natural stones have a size of 0.1–1 mm usually, and large diamonds have a big value in jewelry. Due to these reasons, natural diamond is not considered as a material for high-tech applications.

The main methods for producing synthetic diamonds are the HPHT temperature gradient method [5], chemical vapor deposition (CVD) [6], and detonation synthesis [7]. The latter method is also called shock-wave method and produces the diamond nanopowder. It is economically viable in the field of conventional explosive weapon recycling. This nanopowder is used for seeding substrates for CVD synthesis of diamond [8], in adaptations of quantum key distribution algorithms [9], medical research [10], etc.

HPHT synthesis allows obtaining diamond crystals up to 20 mm in size. Using laser cutting from these stones, diamond substrates with a thickness of 0.25–3 mm are obtained. At present, diamond substrates with dimensions of 15 × 15 mm^2 [11] are commercially available. Unfortunately, large-size HPHT diamonds are characterized by an inhomogeneous distribution of impurities (N, Ni, etc.) in volume due to the inhomogeneous incorporation of impurities during the growing process into various growth sectors [12]. However, the HPHT method enables to synthesize diamond samples in which, using radiation-thermal treatment, it is convenient to create the necessary concentrations of spin- and optically active NV centers, which are intensively studied for applications of quantum cryptography [13], quantum computing [14], spintronics [15], and magnetometry [16]. The very important

IntechOpen

application of HPHT diamonds is a role of substrates for the deposition of high-quality thin films and bulk diamond samples using CVD method.

The priority of the first CVD synthesis patents relates to the 1950s of the last century [17, 18]. But only at the end of the century CVD diamond technology has developed enough to deposit graphite-free epitaxial diamond film on a diamond substrate [19]. At the time being CVD synthesis is a method for producing single-crystal diamonds of the highest quality, superior to the best jewelry brilliants. Depending on the choice of the substrate, the CVD method can produce both single crystals on substrates of diamond [20] and iridium [21], as well as polycrystalline samples on substrates of silicon [22, 23], copper [23], iron [23], etc.

The highest thermal conductivity of diamond (4–5 times higher than that of SiC and copper) enables to divert tens to hundreds of watts of heat generated from electronic devices, which determined the use of this material as a heat sink/radiator for powerful electronic devices [24]. For example, the use of CVD diamond as a substrate for GaN-based heterostructures provides a low thermal resistance at the material interface and the absence of thermal greases [25, 26]. At the same time, the diamond may serve as a passive element itself (heat sink) or as the basis of the active element, simultaneously performing the heat sink function.

The low dielectric constant of diamond means a high propagation velocity of electromagnetic radiation in the material and lower dielectric losses at high frequencies, due to lower parasitic capacitances. This fact additionally makes the use of diamond attractive for developing microwave and extremely high-frequency electronics, optoelectronics, and photonics devices of the future. Such integrated devices will be less energy intensive compared to devices based on traditional semiconductors. These characteristics, combined with high heat dissipation properties, makes diamond very attractive for creating electronic components for avionics of supersonic and hypersonic aircraft and spacecraft, i.e., where the problem of overheating is very important [27].

The size of polycrystalline CVD diamonds already reaches 120 mm in diameter [28]. But they are characterized by a high degree of roughness of the growth surface (0.1–10 μm). The nucleation surface of the CVD diamond substrate is smoother (0.01–0.1 μm) but more defective, which impairs its thermal, electronic, and optical properties. Difficulties in the mechanical processing of diamond led to the development of laser methods for its processing. In addition to laser cutting [29], separating the substrate [30], and polishing the surface [31], laser methods have found their application in the creation of antireflection structures on the surface of the diamond [32], which is important for the problems of radiation extraction from a material with a high refractive index and its significant dispersion (2.40–2.47 in the visible range).

Diamond is characterized by high bipolar mobility of charge carriers and is inferior only to gallium arsenide in electron mobility. In diamond, the mobility of electrons exceeds the mobility of holes by tens of percent, while in other semiconductors this difference can reach tens of times. High mobility of charge carriers in the material indicates a high rate performance of the device based on it. To obtain p-type conductivity, diamond is doped with boron in the deposition process, which is a well-established process from a technological point of view [33]. To obtain n-type conductivity, the diamond is doped with phosphorus [33]. But the quality of the phosphorus-containing diamond layers is significantly inferior to boron-containing ones because of the higher value of mechanical stresses, because phosphorus is usually incorporated into diamond in the form of impurity-defective complexes PV and PV_2 (where V is a vacancy) [34]. To reduce the value of mechanical stresses during the doping diamond with phosphorus, lithium is used as a co-impurity [35].

The electrical properties of diamond are significantly affected by the charge of the surface states of the sample. Depending on the type of chemisorbed elements, the electronic affinity of the diamond surface is changeable [36, 37]. The detection of negative electron affinity of the diamond surface (downward bending of zones, the absence of a potential barrier) determined the studies on the creation of cold cathodes and electron emitters based on diamond coatings [38].

Due to the strong covalent bonds, diamond is a chemically inert material and is suitable for medical and biotechnological applications [10]. In this case, such diamond properties as low cost of nanoparticles, the possibility of surface activation by photochemical methods, high photoluminescence yield and its resistance to photobleaching, low cytotoxicity, and, as a consequence, environmental safety are used.

The collection of scientific chapters proposed to the reader reflects some of the aspects, mentioned above. This collection will be useful to students, graduate students, and researchers who are interested in such disciplines as the synthesis of carbon and superhard materials, carbon electronics and photonics, surface chemistry, etc.

Author details

Evgeniy Lipatov[1,2]* and Dmitrii Genin[1,2]

1 Tomsk State University, Tomsk, Russian Federation

2 Institute of High Current Electronics, Tomsk, Russian Federation

*Address all correspondence to: lipatov@loi.hcei.tsc.ru

References

[1] Tallaire A, Mille V, Brinza O, Thi TNT, Brom JM, Loguinov Y, et al. Thick CVD diamond films grown on high-quality type IIa HPHT diamond substrates from new diamond technology. Diamond and Related Materials. 2017;**77**:146-152

[2] World Diamond Market, Alrosa. Available from: http://eng.alrosa.ru/diamonds-explained/the-world-diamond-market [Accessed: 07 May 2020]

[3] Made-in-China Diamonds Poised to Shape Global Market, Xinhua. 2019. Available from: http://www.china.org.cn/business/2019-02/07/content_74445839.htm [Accessed: 07 May 2020]

[4] Zaitsev AM. Optical Properties of Diamond. Berlin Heidelberg: Springer; 2001

[5] Dobrinets IA, Vins VG, Zaitsev AM. HPHT-Treated Diamonds. In: Springer Series in Materials Science. Vol. 181. Berlin Heidelberg: Springer; 2013

[6] Schwander M, Partes K. A review of diamond synthesis by CVD processes. Diamond and Related Materials. 2011;**20**(9):1287-1301

[7] Danilenko VV. On the history of the discovery of nanodiamond synthesis. Physics of the Solid State. 2004;**46**(4):595-599

[8] Buijnsters JG, Carabineiro SAC, Silva AMT, Pastrana-Martinez LM, Morales-Torres S, Faria JL, et al. The use of nanodiamonds in the seeding of CVD diamond and in heterogeneous catalysis. In: 2015 1st Workshop on Nanotechnology in Instrumentation and Measurement (NANOFIM). IEEE; 2015. DOI: 10.1109/NANOFIM.2015.8425346

[9] Andersen SKH, Kumar S, Bozhevolnyi SI. Ultrabright linearly polarized photon generation from a nitrogen vacancy center in a nanocube dimer antenna. Nano Letters. 2017;**17**(6):3889-3895

[10] Turcheniuk K, Mochalin VN. Biomedical applications of nanodiamond. Nanotechnology. 2017;**28**:252001

[11] New Diamond Technology. Available from: http://ndtcompany.com [Accessed: 07 May 2020]

[12] Babich YV, Feigelson BN, Yelisseyev AP, Chepurov AI. Nitrogen incorporation in octahedral diamonds grown in the Fe-Ni-C system. Geochemistry International. 2012;**50**(2):179-184

[13] Pfender M, Aslam N, Simon P, et al. Protecting a diamond quantum memory by charge state control. Nano Letters. 2017;**17**(10):5931-5937

[14] Santagati R, Gentile AA, Knauer S, Schmitt S, Paesani S, Granade C, et al. Magnetic-field learning using a single electronic spin in diamond with one-photon readout at room temperature. Physical Review X. 2019;**9**(2):021019

[15] Zhukov IV, Anishchik SV, Molin YN. ODMR spectroscopy of NV—Centers in diamond under high MW power. Applied Magnetic Resonance. 2017;**48**(11-12):1461-1469

[16] Anishchik SV, Vins VG, Yelisseyev AP, Lukzen NN, Lavrik NL, Bagryansky VA. Low-field feature in the magnetic spectra of NV-centers in diamond. New Journal of Physics. 2015;**17**:023040

[17] Spitsyn BV, Deryagin BV. The Way to Build up Facets of Diamond. USSR Certificate of Authorship 339134 [Accessed: 10 July 1956]

[18] Eversole WG. Synthesis of Diamond. US Patent 3030187 [Accessed: 23 July 1958]

[19] Nishimori T, Nakano K, Sakamoto H, Takakuwa Y, Kono S. n-Type high-conductive epitaxial diamond film prepared by gas source molecular beam epitaxy with methane and tri-n-butylphosphine. Applied Physics Letters. 1997;**71**(7):945-947

[20] Wang X, Duan P, Cao Z, Liu C, Wang D, Peng Y, et al. Surface morphology of the interface junction of CVD mosaic single-crystal diamond. Materials. 2019;**13**(1):91

[21] Lee KH, Saada S, Arnault JC, Lee KH, Moalla R, Saint-Girons G, et al. Epitaxy of iridium on $SrTiO_3$/Si (001): A promising scalable substrate for diamond heteroepitaxy. Diamond and Related Materials. 2016;**66**:67-76

[22] Mishra SC, Choudhary RK, Mishra P, Abraham GJ. Deposition and characterization of diamond thin films by HF-CVD method. AIP Conference Proceedings. 2015;**1665**:080060

[23] Gautama P, Toyota H, Iwamoto Y, Zhu X, Nomura S, Mukasa S. Synthesizing diamond film on Cu, Fe and Si substrate by in-liquid microwave plasma CVD. Precision Engineering. 2017;**49**:412-420

[24] Hindle P. Diamonds are a high-power engineer's best friend. Microwave Journal. 2018;**61**(6):34-36

[25] Han Y, Lau BL, Tang G, Zhang X. Thermal management of hotspots using diamond heat spreader on Si microcooler for GaN devices. IEEE Transactions on Components, Packaging and Manufacturing Technology. 2015;**5**(12):1740-1746

[26] Chu KK, Chao PC, Diaz JA, et al. S2-T4: Low-temperature substrate bonding technology for high power GaN-on-diamond HEMTs. In: IEEE Lester Eastman Conference on High Performance Devices (LEC-2014). 2014. pp. 1-4

[27] Zhukov S. Complete Upgrading of the MiG-31BM: A Factor That Limited the Capabilities of the High-Altitude Interceptor of the Russian Federation Was Eliminated. 2018. Available from: https://newinform.com/101686-polnaya-pereborka-mig-31bm-ustranen-faktor-ogranichivavshii-vozmozhnosti-vysotnogo-perekhvatchika-rf [Accessed: 07 May 2020] (in Russian)

[28] The Element Six CVD Diamond Handbook. Available from: https://e6cvd.com/media/wysiwyg/pdf/E6_CVD_Diamond_Handbook.pdf [Accessed: 07 May 2020]

[29] Kononenko TV, Ralchenko VG, Ashkinazi EE, Polikarpov M, Ershov P, Kuznetsov S, et al. Fabrication of polycrystalline diamond refractive X-ray lens by femtosecond laser processing. Applied Physics A. 2016;**122**(3):152

[30] Temnov AM. Hole formation in a diamond substrate of hybrid-monolithic integral SHF schemes. Journal of Communications Technology and Electronics. 2020;**65**(2):179-187

[31] Scalbert W, Tanner D, Holtz R. Development of high-power laser ablation process for polycrystalline diamond polishing, part 2: Upscaling of PCD ultra-short pulsed laser ablation to high power. Proceedings of SPIE. 2018;**10525**:105250N

[32] Granados E, Martinez-Calderon M, Gomez M, Rodriguez A, Olaizola SM. Photonic structures in diamond based on femtosecond UV laser induced periodic surface structuring (LIPSS). Optics Express. 2017;**25**(13):15330

[33] Kalish R. Doping of diamond. Carbon. 1999;**37**(5):781-785

[34] Schwingenschlögl U, Chroneos A, Schuster C, Grimes RW. Doping and cluster formation in diamond. Journal of Applied Physics. 2011;**110**(5):056107

[35] Shao QY, Wang GW, Zhang J, Zhu KG. First principles calculation of lithium-phosphorus co-doped diamond. Condensed Matter Physics. 2013;**16**(1):13702

[36] Attrash M, Kuntumalla MK, Michaelson S, Hoffman A. Nitrogen-terminated polycrystalline diamond surfaces by microwave chemical vapor deposition: Thermal stability, chemical states, and electronic structure. The Journal of Physical Chemistry C. 2020;**124**(10):5657-5664

[37] O'Donnell KM, Edmonds MT, Tadich A, Thomsen L, Stacey A, Schenk A, et al. Extremely high negative electron affinity of diamond via magnesium adsorption. Physical Review B. 2015;**92**(3):035303

[38] Romanyuk O, Bartos I, Gordeev I, Artemenko A, Varga M, Izak T, et al. Electron affinity of undoped and boron-doped polycrystalline diamond films. Diamond and Related Materials. 2018;**87**:208-214

Chapter 2

Polycrystalline Diamond Characterisations for High End Technologies

Awadesh Kumar Mallik

Abstract

Characterisations of polycrystalline diamond (PCD) coatings have routinely been done over the past three decades of diamond research, but there is less number of reports on some of its very unique properties. For example, diamond is the hardest known material and, in probing such hard surfaces with any indenter tip, it may lead to damage of the instrument. Due to such chances of experimental accidents, researchers have performed very few attempts in evaluating the mechanical properties of PCDs. In the present work, some of these very special properties of diamond that are less reported in the literature are being re-investigated. PCDs were characterised by photoluminescence (PL), Fourier transform infrared (FTIR) spectroscopy, transmission electron microscope (TEM), and X-ray diffraction (XRD) techniques. The diamond surface was also polished to bring the as-grown micron level of surface roughness (detrimental for wear application) down to few hundreds of nanometer. The tribological properties of such polished and smooth surfaces were found to be appropriate for wear protective coating application. This chapter revisits some of the unreported issues in the synthesis and characterisation of PCD coatings grown on Si wafer by the innovative 915 MHz microwave plasma chemical vapour deposition (MPCVD) technique.

Keywords: CVD diamond, polishing, characterisations, properties

1. Introduction

CVD grown diamond is an important class of material [1–6]. The characteristic of such material is very much dependant on the CVD processing conditions [7–21] and as well as on the post-processing steps that are equally significant for the efficient use of this material in engineering applications [22–26]. There are environments where it is exposed to heat under extreme conditions or it is rubbed against hard ceramic surfaces. Diamond is the hardest known material with very low coefficient of friction which naturally makes it a suitable tribological surface [27–29].

But the as grown diamond surface is very rough in nature which is required to be planarised before putting into real environments. It is well understood nowadays that the diamond nucleation and growth takes place first by coalescence of the seeded layer on the substrate surface and then on top of which diamond crystals grown in columnar fashion. Such vertical growth of crystals causes the top surface to be very rough and simultaneously there may be some inherent porosity present due to rise of such vertical columns. On the other hand, the nucleation side of the freestanding diamond may have some porosity due to random coalescence of the islands during CVD processing. There is no such report of studying the surface area and porosity of CVD grown polycrystalline diamond (PCD). Here, it has been attempted for the first time to evaluate the pore size that may be present in the diamond coatings grown by CVD, using Brunauer-Emmett-Teller (BET) technique.

Moreover, tribological action may cause rise of temperature under humid atmospheric condition for diamond material to degrade. So it is necessary for the CVD grown diamond to be well characterised for the wear and friction applications [30–34]. Thermal stability has been studied so far by many authors [35–37] but it lacks proper assessment of oxidation temperature [38–40] since only standard furnaces have been used so far. Here, thermo-gravimetric analysis (TGA) and differential scanning calorimeter (DSC) techniques have been used for the first time ever to pinpoint the augmentation of oxidation.

Methane or other precursor gas compositions [41–45] determine the defects present inside the diamond crystals. For example, high methane concentration may cause many CVD growth defects which render the grown coating to become black/grey; whereas the cleaner processing conditions of low methane percentages etc. may cause the diamond coating to be white or transparent. Similarly boron or nitrogen in the precursor gases [46] make the diamond blue [47] or yellow [48–50] in colour due to the substitutional atomic defects [51, 52] introduced in the diamond lattice [53, 54]. Such substitutional defects are studied by [55] photoluminescence (PL) [56–58], Fourier transform infra-red (FTIR) spectroscopy techniques [59], whereas the point, line, plane or volume defects during CVD growth can be viewed under transmission electron microscopes (TEM) [60–68]. Post-processing steps such as annealing [69, 70] may be required to remove some of these defects [71]. Annealing [72–76] also makes the diamond to become purer in respect of graphitic inclusions. Raman spectra [77–79] reveal the phase purity of CVD grown diamonds [80–86]. Polishing is routinely done to reduce as-grown diamond surface roughness [87]. Tribology of such polished surfaces has been studied in detail but there is lack of literature [88–90] against silicon nitride ceramics [91] under machine oil lubrication [92]. Moreover, due to very hard and rough top surface of CVD grown PCD, researchers are hesitant to probe the as-grown PCD surface with their expensive Berkovich or other indenters [93–95]. In effect, there is not much literature [96–100] available which describes the mechanical properties [101, 102] of CVD grown PCDs.

Characterisation [103–107] of the CVD grown PCDs is very essential besides its synthesis. Some techniques which are frequently used for powder samples like, BET surface area analysis, TGA-DSC, Zetasizer particle size measurements are being reported here for the first time, for CVD polycrystalline diamond coatings. Moreover, black and white grades of diamond have been grown [108–110] and characterised so far, but their corresponding growth defects have not frequently been reported [111–113]. Here, authors have tried to elucidate the CVD growth defects that are present in their CVD grown diamonds. Afterwards, the PCD samples were polished to make them effective for tribological applications. Later on, such polished diamond surfaces have been mechanically characterised for studying tribological action [114] against silicon nitride balls under machine oil lubricants for the first time.

2. Materials and methods

2.1 Processing of polycrystalline diamond (PCD)

Polycrystalline diamond samples were fabricated with CSIR-CGCRI based DT1800 microwave plasma enhanced CVD (MPCVD) reactor on single crystal Si wafers [115]. Deposition was carried out for 4–10 days to grow 0.5–0.8 mm thick diamond coatings. Reactor pressure and temperature were maintained in between 110 and 120 Torr and 900–1100°C, respectively. 1–3% methane in hydrogen gas mixture was used for growing diamond coatings with 9 kW input microwave power. Afterwards, PCDs were made freestanding by wet chemical etching of Si wafers with 1:1:1 ratio solution of $HF:HNO_3:CH_3COOH$ [116]. Such freestanding diamond wafers were cut in 6 mm diameter smaller discs by laser cutter (Hallmark Plus model. Nd-YAG lamp pumped laser with 10 watt power at 1064 nm wavelength). The diamond sample was then hot mounted to polish the as-grown rough surface [117]. Mechanical polishing was done using Leco, Germany polisher and subsequently CP4 model, Bruker, USA machine was used for chemo-mechanical polishing of the PCD samples. Metal bonded diamond disc and Cu-bonded diamond discs were used with different grit sizes for mechanical polishing with water as cooling agent. On the other hand, the chemo-mechanical polishing of diamond samples was carried out against alumina ceramic disc using $K_2S_2O_8$ solution and concentrated sulphuric acid along with diamond pastes. Separately, annealing of few diamond samples was also done at 600°C for 1 h in standard air furnace.

2.2 Physical characterisations of PCDs

The detonation nanodiamond (DND) particle size of the seeding slurry was measured by Malvern make Zetasizer instrument. Brunauer-Emmett-Teller (BET) surface area analyser (Quantachrome Instruments) was used to measure the surface area; and nitrogen adsorption isotherm (Barrett-Joyner-Halenda or BJH method) was used to measure the pore size and volume in the freestanding diamond coatings. The samples were degassed under vacuum at 100°C for 1 h prior to measurement. Thermogravimetric (TGA) and differential scanning calorimetry (DSC) analysis of the freestanding diamond sample was done at a heating rate of 10°C min^{-1} on a simultaneous thermal analyser (STA 449F, Netzsch, Germany). X-ray diffractograms (XRD) were recorded in the 2θ range 30°–100° at a slow scanning rate of 1° min^{-1} by an X-ray diffractometer (Philips X'Pert, The Netherlands) with Cu Kα radiation (at 40 kV and 40 mA). Transmission electron microscopy (TEM-Tecnai G2 30ST, FEI Company, USA) was used to evaluate the defects present inside the diamond lattice. FTIR/FIR spectrophotometer (Model: Frontier IRL 1280119, Perkin Elmer) attached with reflectance measurement arrangement with incident and reflectance angles set at 22.5°, was used for studying opaque and translucent PCD samples. The photoluminescence (PL) studies including emission and excitation measurements were carried out on a fluorescence spectrophotometer (Model: Quantum Master, enhanced NIR, from Photon Technologies International) using a xenon arc lamp of 60 W as a pump source. Raman spectra were obtained using a STR500, Cornes Technologies, (formerly known as Seki Technotron) micro-Raman spectrometer, with excitation by argon ion (514.5 nm) laser. The resolution was about 1 cm^{-1} with 1200 grating size. In all the Raman experiments, the laser spot size was 1–2 μm and an exposure time of 20 s was used which was repeated thrice to acquire the Raman signals during each measurement.

2.3 Mechanical and tribological characterisations of PCDs

Polished surfaces of the PCD samples were seen under a ZEISS Supra 35 (Germany) field emission scanning electron microscope (FESEM) with EDAX attachment. Roughness of the PCD surfaces were evaluated by a contact profilometer (Talysurf PGI 2000S, Taylor Hobson). The nanohardness and Young's modulus were measured by nanoindentation technique applied to the plan section of the CVD grown polycrystalline diamond polished and nucleation side surfaces. A typical load of 1000 mN was used for the nanoindentation experiments. A standard nanoindenter (Fischerscope $H100XY_P$, Fischer, Switzerland) was used for this purpose. The nanoindenter machine was operated according to the DIN 50359-1 standard. It offered a load range of 0.4–1000 mN. It was equipped with a Berkovich indenter. The indenter possessed a tip radius of 150 nm. The machine was used in ambient laboratory conditions (23 ± 4°C, 70 ± 5% relative humidity). The depth sensing resolution of the machine was 1 nm. The force-sensing resolution of the machine was 0.2 μN. The machine was calibrated with nanoindentation based independent evaluation of H ≈ 4.14 ± 0.1 GPa and E ≈ 84.6 ± 3.5 GPa of a BK7 Glass (Schott, Germany). This material was provided by the supplier as a standard reference block. A ball-on-disc tribometer assembly (NANOVEA) was used for highly accurate and repeatable wear friction testing in rotational modes. The counter-face was a Si_3N_4 ball which was attached to a load arm and the load arm was directly in contact with experimental PCD disc surface. 3 mm diameter Si_3N_4 ball was in circular motion with 5 N applied normal load against PCD surface using commercially available machine oil lubricant in ambient laboratory condition. The speed of the ball was 0.04 m/s. Optical microscope (Olympus BX 51, country) study was done after each tribological experiment to know the radius of worn out silicon nitride ball.

3. Results and discussion

3.1 Physical characterisations—DND seed particle size, PCD crystallinity, porosity, growth defects and thermal stability

Detonation nanodiamond (DND) suspension in dimethyl sulfoxide (DMSO) by 0.5 wt% is used as seeding slurry for CVD growth of polycrystalline diamond material [118]. Nucleation density determines the faster grain coalescence phenomenon during chemical vapour deposition. It has been observed that DND seeds are much superior in enhancing nucleation of diamond than conventional micron size grits for seeding diamond nuclei on the foreign substrates [119]. The particle size of seeds is thus important in determining nucleation efficiency. One commercially available DND in DMSO was used for seeding. Before using to make seeding suspension by mixing it with 3 parts of methanol, it was essential to know the exact particle size distribution of such suspensions. Dynamic light scattering (DLS), is a non-invasive, well-established technique for measuring the size and size distribution of molecules and particles. The Brownian motion of particles or molecules in suspension causes laser light to be scattered at different intensities. Analysis of these intensity fluctuations yields the velocity of the Brownian motion and hence the particle size using the Stokes-Einstein relationship. The DND seeding suspension was procured from a commercial supplier and it was being used from time to time over a period of 1 year. When the particle size was calculated from DLS technique using Malvern make Zetasizer, it was found that the original suspension has become agglomerated considerably. The fresh stock of commercial suspension came with 25–30 nm DND particle size on its label but to our surprise after one long year in the shelf, the suspension only has 7% with original size of 25.56 nm whereas, 93% of the DND particles agglomerated to give an average size of 123.4 nm, as shown in the following **Figure 1a**. But such agglomeration does not appear to hinder its seeding efficiency.

The as-grown surfaces of the PCD coatings are very rough in nature which is labelled in **Figure 1b**. Cross-sectional microscopy image reveals columnar

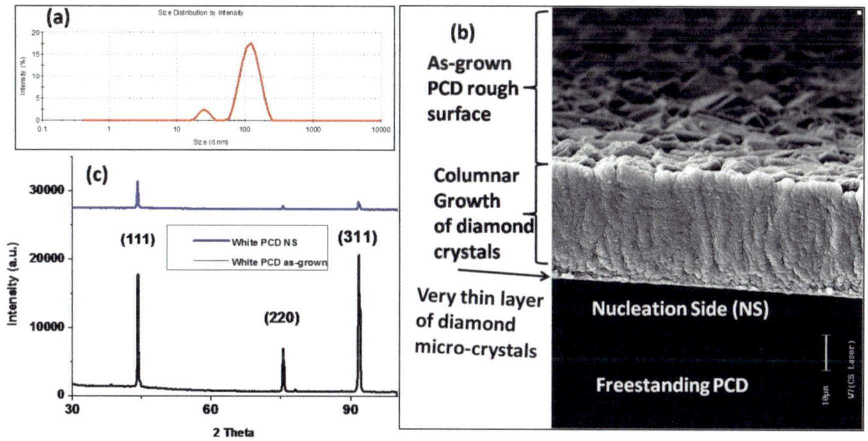

Figure 1.
(a) DND-DMSO seeding suspension particle size distribution; (b) cross-sectional SEM image of freestanding PCD; and (c) corresponding XRD signals from as-grown and nucleation side (NS).

growth of diamond crystals from the substrate side. The interface between the Si wafer and diamond columns consists of a very thin layer of diamond micro-crystals [116] which were initially coalesced with each other to form a continuous coating before diamond could grow vertically on top of them. Such thin crust layer on the nucleation side gives very poor XRD signals (top of **Figure 1c**) corresponding to diamond reflection planes, whereas, the peak intensities from the opposite as-grown surface (bottom one of **Figure 1c**) is very strong and characteristics of diamond cubic crystals. DND seeds initially formed islands which grow in a single layer until touching upon each other to form a continuous film. The diamond columns also do not grow exactly at 90° in the z direction. The adatoms from the plasma may add in a random way to produce some misorientation between the adjacent diamond columns. Moreover the top surface of diamond PCDs are very rough and sometimes appear porous. Such, (i) random coalescence of thin crust layer on the nucleation side, (ii) misorientation in between diamond columns and (iii) uneven as-grown surface of the diamond films, suggest that the CVD grown diamond may have some pores present in the freestanding coating. It is also true that in practice we assume that the CVD grown films are having 100% theoretical density but it is understood that that is not actually the case. So in order to know the pores that might present, BET surface area was calculated for the PCDs. Gas molecules are dosed into the sample chamber after cooling the solid down to a constant temperature in order to partially accumulate the gas molecules onto the solid surface which can be applied for the characterisation of surfaces as well as for the pore characterisation. The nitrogen adsorption at the temperature of liquid nitrogen (77 K) is standard method for such surface characterisation [120]. It is assumed that the nitrogen condenses onto the surface in a monolayer and so once the size of the gas atoms/molecules are known, the surface area can be estimated from the amount of adsorbed gas.

It was found that white PCD is having less surface area compare to the black PCD (**Table 1**). Nitrogen adsorption and desorption isotherms measured at 77 K allow the determination of pore volume and sizes. It was similarly observed that both the values are higher for black colour PCD. So it is inferred that black PCDs are having more CVD growth defects which is giving bigger pores and higher amount of porosity—i.e. addition of carbon atoms into the diamond sp3 lattice is more incoherent.

PCD sample	Surface area (m^2/g)	Pore volume (10^{-3} cc/g)	Pore diameter (nm)
White	1.19	1.82	3.51
Black	4.17	6.74	3.72

Table 1.
Freestanding PCD surface area and pore analysis.

PCDs grown using 3% CH_4 in H_2 resulted in black colour (**Figure 2a**), whereas, the coatings were white translucent in nature (**Figure 2b**) when 1% CH_4 was used. Colour of the freestanding diamond coatings vary due to the presence of defects inside the diamond lattice. For example when boron or nitrogen is substituting the carbon atoms, it results in blue or yellow colour of the diamond crystals. The PCD could become completely opaque or black in colour due to the presence growth defects like dislocations, twining, grain boundaries, stacking faults etc. [121]. **Figure 2c** and **2d** shows transmission electron microscopy images of the line defects pinned at grain boundary in one such poorer variety of black diamonds grown using 3% methane in hydrogen. On the other hand when 1% CH_4 was used to deposit diamond, the crystals were less defective and could synthesise white transparent freestanding coatings as shown in **Figure 2e**.

The quality of white PCD sample was found to be very high by Raman Spectroscopy. It has been found that under atmospheric conditions, the CVD diamond sample could sustain up to 700°C furnace temperature, but 750°C was not tolerable for the diamond material to resist oxidation. As per available literature [38], the procedure to study the thermal stability is to heat the diamond sample in an environment controlled furnace, but such procedure does not pinpoint the oxidation temperature of diamond material. In order to know the exact oxidation temperature of the CVD diamond prepared by DT1800 reactor, it was decided to conduct TGA-DSC experiments.

Figure 3 shows the TGA curve for white PCD sample when heated from room temperature at constant heating rate of 10 K/min in air with alumina crucible up

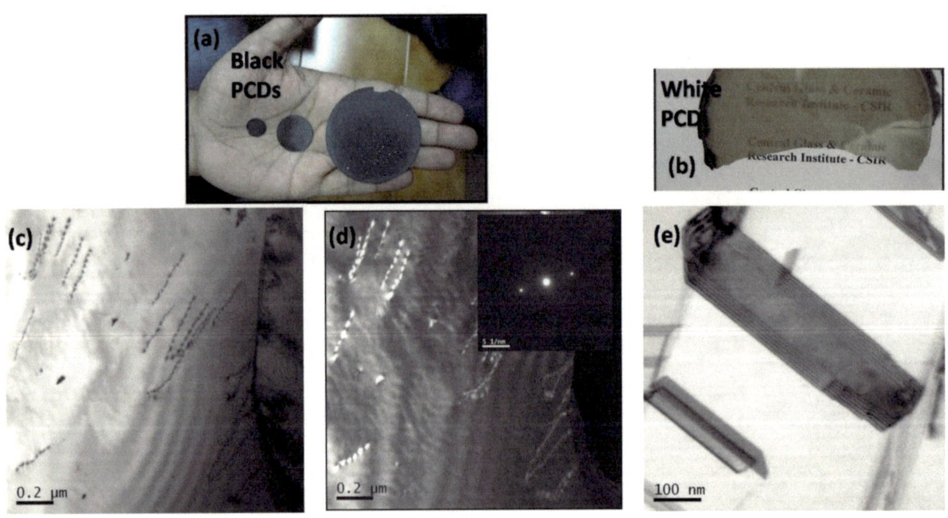

Figure 2.
(a) Black grade PCDs, (b) white grade PCD, (c) bright-field TEM image of a black PCD grain with some dislocations pinned at the grain boundary. (d) Corresponding wide beam dark field (WBDF) image showing clearly discernible dislocations. Inset shows the two-beam condition used for imaging the dislocations, (e) bright-field TEM image of defect free White PCD grain with some twins and grain—boundaries.

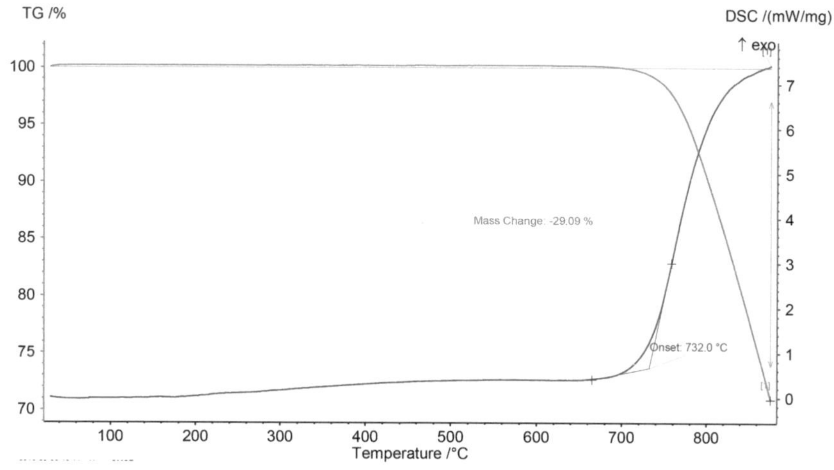

Figure 3.
TGA and DSC curves for heating CVD diamond in air.

to 900°C. There was no change in weight until 700°C temperature was crossed, after which there was continuous drop in weight, losing 29.09% of the initial value. The corresponding DSC curve shows that the exact onset temperature for oxidation reaction is 732°C, after drawing tangents. The weight loss can be attributed to the loss of carbon from diamond as gaseous by product of oxidation reaction. So, further to study the behaviour of CVD grown defects in PCD samples, they were heated in an air furnace at 600°C for 1 h. Such TGA-DSC curves confirm that annealed PCD samples do not undergo degradation on annealing and are good for studying the effect of heating on their internal defects.

3.2 Spectroscopy: FTIR, PL and Raman—PCD defects and annealing effects

3.2.1 FTIR studies

The FTIR reflectance spectra of as grown PCD samples, white PCD and black PCD, together with white PCD sample after being annealed at 600°C for 1 h are presented in **Figure 4**. The spectra reveal the presence of a broad absorption peak in the range 1745–2840 cm^{-1} in white PCD sample which may be attributed to the characteristic two phonon absorption of diamond [122, 123] and it is seen to be almost absent in black PCD sample. However, this peak disappears completely in annealed white PCD sample with simultaneous appearance of a new peak at 1090 cm^{-1}. Interestingly, the peak at 1090 cm^{-1} is observed to be more intense in as prepared black PCD sample. This peak marks the presence of N-defects in diamond lattice [123]. Thus, it can be stated that the as grown white PCD sample contains less nitrogen defects as compared to black PCD and is of better quality in terms of non-diamond inclusions. As a result characteristic carbon-carbon bond absorption vibrational modes are more distinct in white PCD as compared to black PCD, which has many internal planar and volume defects as already observed in **Figure 2**. But, annealing at 600°C results in the growth of atomic defect centres in white PCD sample. As the nitrogen impurity centres grow at the expense of carbon replacement in the lattice, intrinsic absorption band due to diamond is weakened while the vibrational band due to N-defects intensifies.

In addition, a third distinct peak is noticed at around 1490 cm^{-1} in both as grown white PCD and annealed sample which may be assigned to absorption due

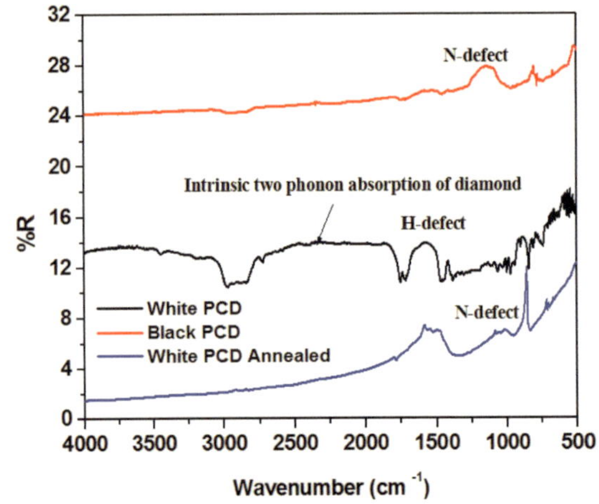

Figure 4.
FTIR spectra of different PCD samples.

to H-centres trapped in the lattice [124]. In black PCD sample this peak is seen to be extremely weak in nature. This may be explained on the basis of the processing condition of the two samples. White PCD was grown with 1% methane and 99% hydrogen whereas black PCD was grown with 3% methane and 97% hydrogen. Thus use of more hydrogen precursor during the CVD process may have resulted in trapping of H-centres in the lattice of white PCD. For further confirmation of the presence of impurity centres, photoluminescence study of these samples was performed.

3.2.2 Photoluminescence studies

Photoluminescence is a characteristic property of several defect centres in lattice and is thus a confirmatory tool for the presence of defects especially in diamonds. In an attempt to have detailed information regarding non-diamond inclusions, the photoluminescence spectra of all the samples were recorded under excitation of 270 nm wavelength and the spectra is presented in **Figure 5**.

The spectra reveal the presence of seven distinct peaks at 470, 496, 567, 625, 689, 734 and 767 nm respectively. Pezzagna et al. [125] suggests that small PL signal for neutral nitrogen vacancy centres is present at 575.67 nm, but for the negatively charged nitrogen vacancy centre, the PL peaks are observed at 637 nm with multiple phonon replica thereafter. It is also well known that silicon vacancy centre emits strong PL signal at 738 nm and a weak replica at 758 nm [121, 126]. Although nitrogen vacancy centres are best photo emitters for diamond but it has strong photon-phonon interactions which is responsible for broad emission spectrum from 600 to 850 nm. That may be the reason for us not getting the nitrogen vacancy signal exactly at the theoretical values in **Figure 5**. The peaks at the 567, 625 and 689 nm can be attributed to nitrogen vacancy centres, but not definitively. Similarly the peaks at 734 and 767 nm in **Figure 5** can be assigned to silicon vacancy centres but again, it is a complex defect centre with an impurity atom and vacancy—so the peaks are not always reproducible at identical positions. Surprisingly there are two more peaks in **Figure 5** at 470 and 496 nm which are typically found for Ni-related defect centres—but usually reported for HPHT synthetic diamonds [127]. It is not yet known about the origin of such PL peaks in CVD grown diamonds. Although the presence of the peaks can be identified in all the three samples, variation in the peak intensity indicates the difference in

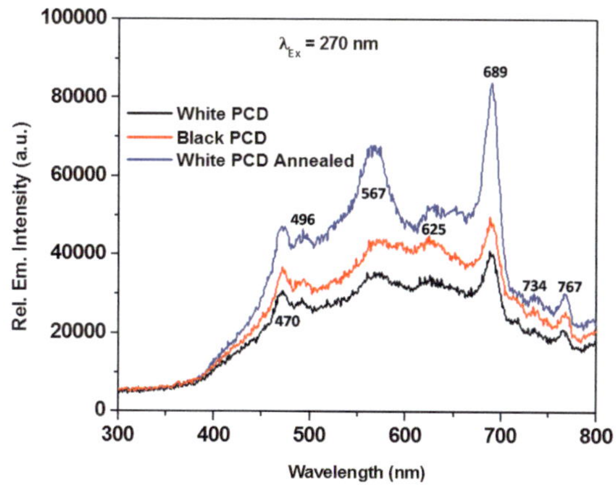

Figure 5.
Photoluminescence spectra of PCD samples.

quality of the samples. It is observed that black PCD sample exhibits the stronger peak intensity as compared to that of white PCD. This indicates white PCD contains lesser defect centres and is of better quality as compared to black PCD which is in accordance with the FTIR results. However, upon annealing white PCD sample is seen to exhibit strongest peak as compared to as prepared white PCD and black PCD samples. This signifies that the concentration of defects in white PCD is enhanced upon annealing which may be explained as due to possible aggregation of defect centres. At elevated temperature, the vacancy in a lattice becomes mobile and moves through the crystal and aggregates [128]. Vacancies upon encounter with isolated impurity atoms (nitrogen, silicon), forms the aggregation of defect centres thus resulting in enhanced luminescence.

3.2.3 Raman spectroscopy

The diamond peak positions of both the white and black annealed PCD samples are at 1328.45 cm^{-1} which is a downshift from the theoretical peak position for single crystal diamond as shown in **Figure 6**. Down shift of spectra is due to the tensile stress inside the lattice (thermal stress due to expansion mismatch with the Si substrate does not exist as the PCDs are freestanding) [129]. Now on the other hand, a-C (amorphous carbon) hump centre is around 1486 cm^{-1} for white annealed PCD but such diamond like carbon (DLC) phase centre shifts to 1560 cm^{-1} for the black annealed PCD sample. So this a-C hump central position change may be an indication of the nature of predominant amorphous carbon phase that is present in the samples. It is essentially trans-polyacetylene (TPA) for white annealed sample; whereas, the non-diamond carbon inside black annealed PCD is primarily graphitic carbon. The ratio of peak intensities of diamond to DLC gives estimation about the quality of the deposited diamond coating. Now to calculate the area under the curve, corresponding to the sp3 phase, the integration has been carried out from 1300 to 1360 cm^{-1}, which therefore would take into account of any upshift/downshift of the diamond peaks and disordered graphite that might also contaminate the coating. The integrated area for DLC phase is calculated from 1450 to 1600 cm^{-1}. As expected the quality of white annealed diamond (41%) has found to be better than the quality of the black annealed PCD (33%) sample. The FWHM value of diamond peak is also better for the white annealed PCD (4.62 cm^{-1}) than the black annealed PCD (7.38 cm^{-1}), as found from the analysis of the curves in **Figure 6**.

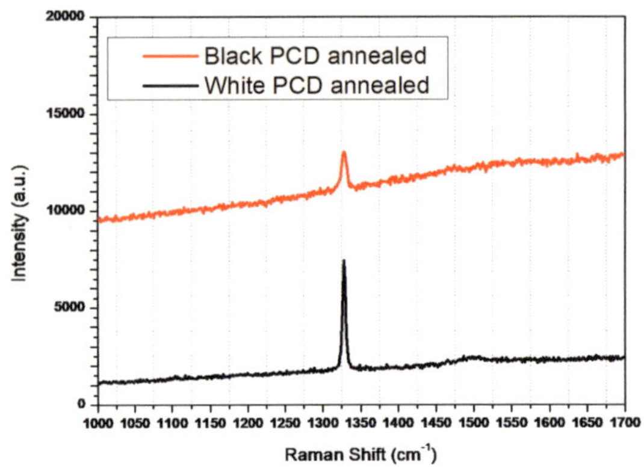

Figure 6.
Raman spectra of freestanding annealed PCD samples.

3.3 Mechanical characterisations: PCD polishing, nanoindentation and tribology

3.3.1 Polished PCD surfaces

Annealed samples were different from the PCD samples that were polished for mechanical property and tribology studies. Polishing was done to reduce the as grown surface roughness (**Figures 2b** and **7a**) which were in the order of 3–4 μm. Before starting polishing experiments, the white as grown PCDs were characterised by Raman spectroscopy (**Figure 7b**). It was found that the as-grown diamond samples were having tensile stress of about 0.8 GPa (calculated from the peak shift data). It also had some non-diamond impurity in the form of TPA which gives very little undulations around 1508 cm^{-1}. The FWHM was calculated to be 5.5 cm^{-1} at the diamond peak position. But when the Raman spectra were taken from the same white as-grown PCD sample after polishing, there is considerable enhancement of diamond peak intensity with appearance of graphitic D peak—giving shoulder to the sp3 peak position, centred around 1329 cm^{-1}. Tensile stress appears to increase on the surface of the same white PCD sample up to 1.34 GPa after polishing action. Moreover, the a-C hump becomes apparent in the polished sample which was not prominently present in the un-polished white PCD sample. The centre of a-C hump is around 1492 cm^{-1}, which indicates that the non-diamond composition was essentially trans-polyacetylene (TPA). FWHM of the diamond peak increases for the polished sample (7.65 cm^{-1}) which also confirms the degradation of the quality of the original (55%) white PCD surfaces after polishing (26%). **Figure 7c** shows the EDAX signals received from such polished surfaces under electron microscope. It can be seen that the carbon peak is not predominant due its low atomic weight. There are prominent peaks corresponding to metal elements like Fe, Bi, which might came from the metal bonded polishing discs. Moreover there are oxygen and sulphur elemental peaks also present, which are essentially from the oxidising chemicals used during chemo-mechanical polishing.

In order to reveal the morphology of polished diamond surfaces, **Figure 8** is provided with successive higher magnification images up to 25kX. It can be seen that there are longitudinal polishing marks, as shown by double headed arrows. Essentially the area under observation is covered with wear debris generated during both mechanical and chemo-mechanical polishing. Due to abrasive action of the polishing materials, fine particles were generated, which, when reacted with the polishing chemicals, were oxidised to dislodge diamond particles from

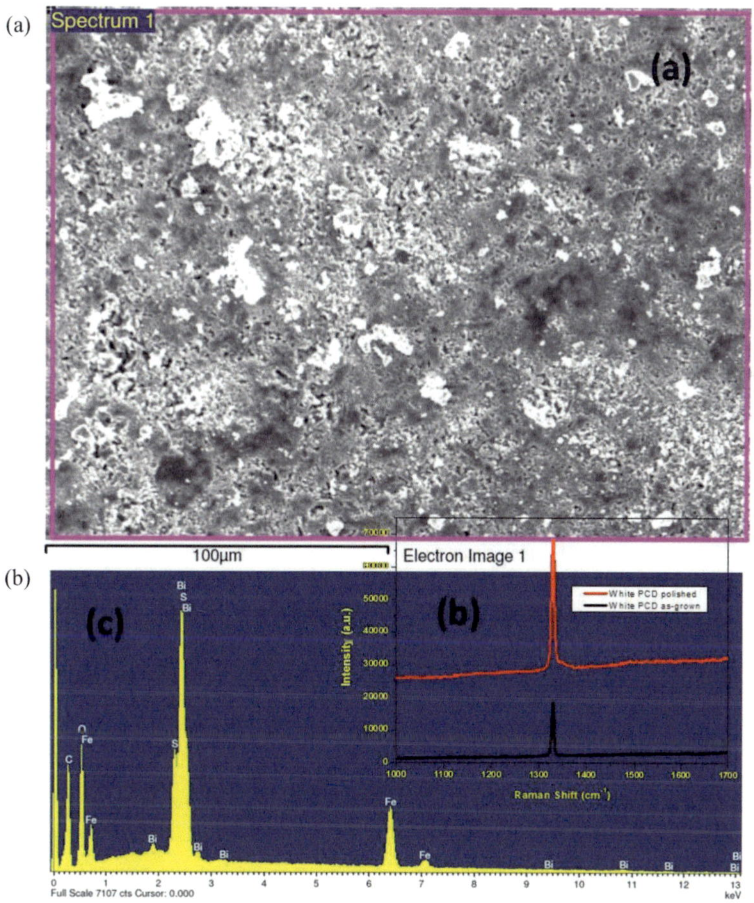

Figure 7.
(a) area scan of polished white PCD sample; (b) Raman spectra of as-grown and polished PCD samples; (c) EDAX signals from the scan area in (a) of the polished PCD sample.

the as-grown rough surfaces. Thus the final microstructures were like wear debris compressed under polishing load with intermediate pores and cracks. Some regions were very smooth with grooves present inside, whereas, their adjacent areas were very rough with scattered particles spread all over. It appears that such unclean top surface after polishing has to be annealed/chemically cleaned in order to remove non-diamond impurities and wear debris, before putting into use for technological application.

3.3.2 Hardness and elastic modulus

Nanoindentation was carried out on black and white PCD samples, both on the polished surfaces and on the mirror smooth nucleation sides (NS). Altogether four PCD surfaces were indented, as shown in **Figure 9**. The roughness of each surfaces are labelled in each **Figure 9**.

At least 10 nanoindentations were made along a line for the plan sections of each CVD grown polycrystalline diamond surfaces, at an applied load of 1000 mN. The time to reach the peak load was kept at 30 s for all the nanoindentation experiments. The unloading time was also kept same as the loading time. The experimental conditions for all the nanoindentation measurements were kept identical. From

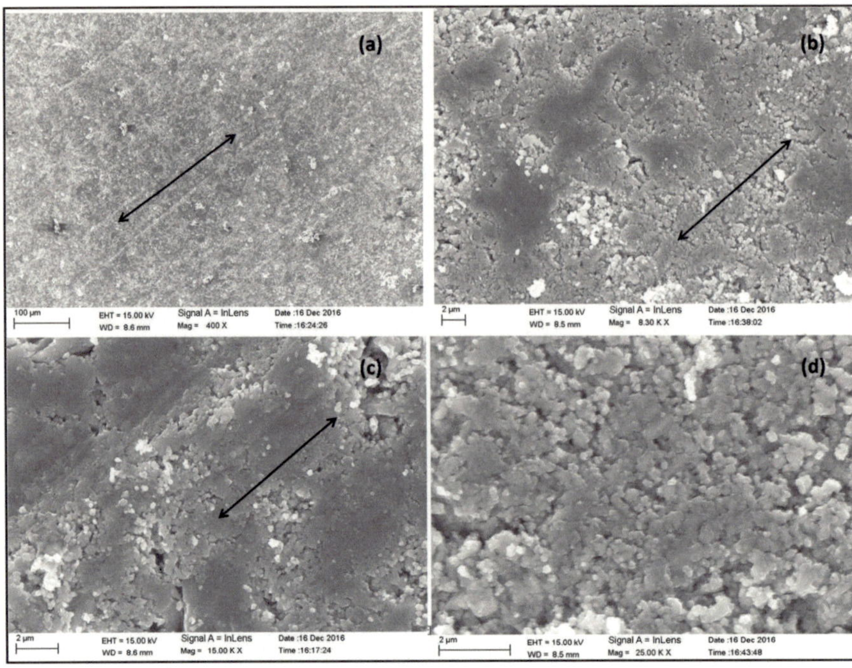

Figure 8.
SEM images of unclean polished white PCD sample at successive higher magnifications.

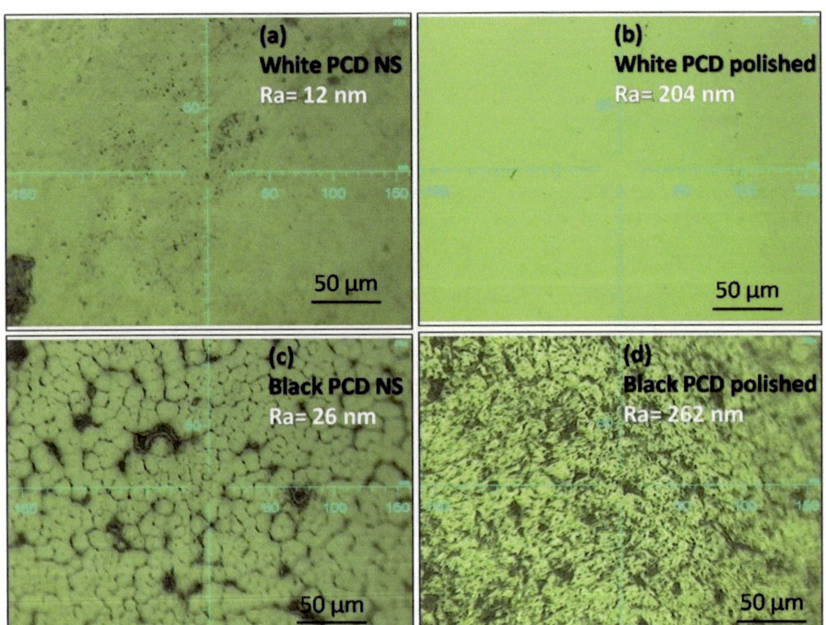

Figure 9.
Four different PCD surfaces probed with nanoindenter.

the experimentally measured load (P) versus depth of penetration (h) data, the nanohardness and Young's modulus values are evaluated for all PCD samples. Oliver and Pharr method [130] was utilised to evaluate the nanohardness and Young's modulus values.

The experimentally measured load (P) versus depth (h) plots for the plan sections of four different types of CVD grown polycrystalline diamond surface

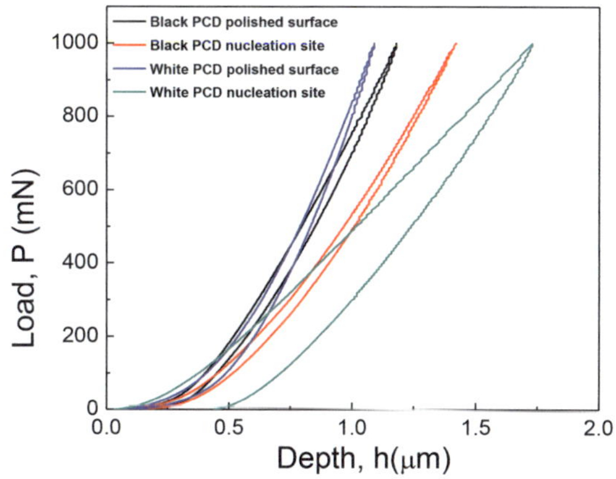

Figure 10.
Load vs. depth plots of nanoindentation PCD samples.

samples are shown in **Figure 10**. In general, all the P-h plots were smooth in nature. An important observation from these data was that at comparable loads, the P-h plots of Black PCD NS, White PCD polished and Black PCD polished were much steeper than those of the White PCD NS sample. Moreover, at comparable loads the depth of penetration data recorded for the Black PCD NS, White PCD polished and Black PCD polished samples was much smaller than those of the White PCD NS sample. These data (**Figure 10**) indicated that nanohardness of Black PCD NS, White PCD polished and Black PCD polished would be higher than that of White PCD NS samples. The P-h plots obtained for White PCD NS samples showed more of the plastic deformation behaviour (**Figure 10**) compared to Black PCD NS, White PCD polished and Black PCD polished. However, the P-h plots obtained for Black PCD NS, White PCD polished and Black PCD polished samples showed more of the elasto-plastic deformation behaviour.

The nanohardness (H) of the PCD samples is shown in **Table 2**. Results show that the polished PCD surfaces have much higher hardness compare to their corresponding nucleation side, as expected. Black PCD has almost 5 GPa less hardness than white PCD polished surface, may be due to the presence of softer carbon phases other than diamond. But on the other hand, their corresponding nucleation sides are showing opposite trend—black PCD NS being harder than White PCD NS. It is already understood from our previous results and discussion that the black PCDs are having more CVD growth defects than white PCDs. Presence of such defects make the poorer variety diamond coatings black or grey in colour; their Raman signals become different as well as their other spectral data. **Figure 1b** and **c** shows that the as grown side is different from the nucleation side. Nucleation side has thin crust like layer whereas the as-grown side is having extended ends of the elongated diamond columnar crystals. From nanoindentation data, it is being revealed that the nucleation side of black quality diamond is harder which may be because of the presence of harder diamond like carbon phases than their white variety. Corresponding average Young's modulus (E) values showed the similar trend as nanohardness. The samples can be ranked in the increasing order of elastic modulus as follows: White PCD NS > Black PCD NS > Black PCD polished> White PCD polished. The hardness and elastic modulus results found in the present work are in agreement with the literature [93, 105] values.

3.3.3 Tribological properties

In order to understand the efficacy and necessity of the polishing method adopted in the present work, the white PCD sample was chosen for further

Samples	Hardness (GPa)	Young's modulus (GPa)
White PCD NS	10.82 ± 0.39	231.72 ± 8.24
White PCD polished	28.50 ± 2.78	1000.06 ± 157.61
Black PCD NS	14.71 ± 2.70	350.32 ± 84.79
Black PCD polished	23.17 ± 4.04	719.72 ± 189.07

Table 2.
Mechanical properties of PCD samples.

tribological studies. Both the polished and the corresponding opposite nucleation side were made to rotate against 3 mm diameter Si_3N_4 counter-face ball under machine oil lubrication. The coefficient of friction was found to be very low in compare to our earlier study of the similar polished/unpolished diamond surfaces against alumina balls [131]. Here also the diamond surfaces do not wear, but the counter body Si_3N_4 wears out, although not as severe as alumina ball which worn out under dry and SBF lubricating conditions by travelling half of the distance travelled by Si_3N_4 balls in the present case. The polished side of white PCD sample is rougher than their corresponding nucleation side (**Figure 9a** and **b**). So, as expected the COF values were found to be higher (~0.08) for polished white PCD surface than the nucleation side of white PCD sample (~0.03). The Hertzian contact pressure was calculated to be 26.2 GPa when 3 mm diameter ball was made to rotate against PCD disc surfaces under 5 N normal load as shown in schematic **Figure 11a**. The worn out ball was investigated under optical microscope (**Figure 12**) and the wear data was calculated as described in authors previous work [131]. It was found that the wear rate for the polished white PCD surface is very small 4.08×10^{-7} mm^3/N m and the wear rate for its corresponding nucleation side is much higher 3.95×10^{-5} mm^3/N m. The hardness of the polished side was found ~28 GPa, whereas its nucleation side was much softer with ~10 GPa nano-hardness. It is counter-intuitive that the counterbody Si_3N_4 wears out more against the softer diamond nucleation surface than while rubbing against the hard polished surface. In order to explain this apparent anomaly, it was necessary to know the hardness of Si_3N_4 from commercial supplier's database and it was found to be 10–12 GPa. Such comparable hardness values of diamond nucleation surface and Si_3N_4 counterbody might have caused nanodiamonds that are present in loosely held colonies [118] on the nucleation side to come out under abrasive wear action of silicon nitride. Such nanodiamonds that come out from the soft nucleation side under abrasive wear action may cause the Si_3N_4 counterbody to further wear out severely, since nanodiamonds are very hard particles.

CVD diamond surfaces sliding against silicon nitride balls reportedly give high COF values under dry lubricating conditions [91]. Researchers have tried commercial soybean-derived biodiesel lubricant [132] to improve the performance of the combustion engine. They could achieve 0.07 as the minimum value of COF when trying self-mated diamond coated Si_3N_4 material. They used 86 m of sliding distance under 2.5–4.5 GPa initial Hertzian contact pressures. Recently a grapheme coating has also been applied on top of microcrystalline diamond surfaces to improve the COF values against Si_3N_4 balls under reciprocating actions [133]. 0.04 was the minimum achievable COF value. The researchers could notice wear track on such grapheme coated diamond flat surfaces. The calculated wear rate of the counterface Si_3N_4 ball from the measurement of the ball scar diameter was found to be in the order of 10^{-5} mm^3/N m. This same group of researchers also conducted a tribo-map of diamond films [134] against silicon nitride ceramics sliding in air, but under reciprocating action (present paper reports circular motion). The normal

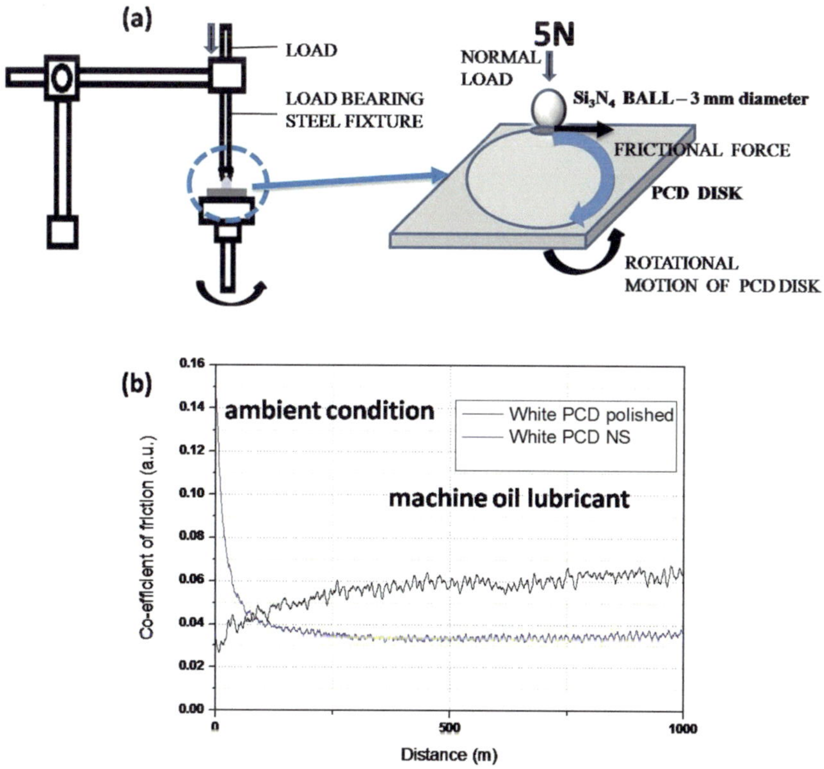

Figure 11.
(a) Schematic of ball-on-disc tribometer and (b) corresponding friction curves.

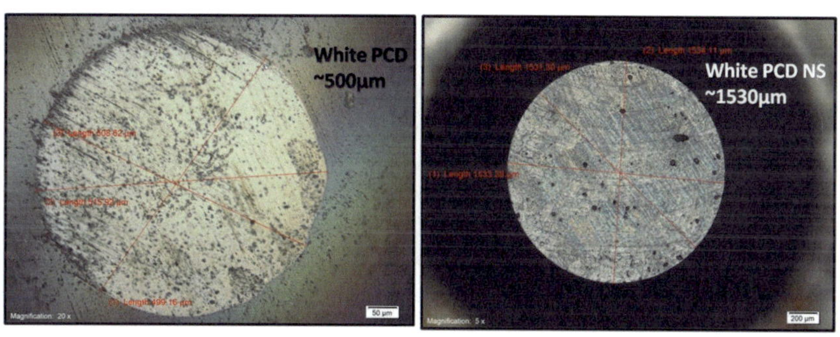

Figure 12.
Wear scar diameters on counterbody Si_3N_4 ball against white PCD (a) polished surface and (b) nucleation side surface.

load was 2–6 N, sliding speed was 0.13–0.3 m/s. They reported 0.07 as the best possible COF value with nanocrytsalline diamond, which was increased to 0.11, when the diamond film became microcrystalline. They also could observe worn out diamond film surface; and profilometer scan wear volume data revealed wear rate less than 10^{-7} mm^3/N m. Bogatov et al. [135] also reported >10^{-7} mm^3/N m wear rate of diamond film sliding against silicon nitride ball (ball characteristics were same as used in the present work) with 0.12 COF after 24,000 sliding cycles. But in the our present experiments with diamond disc, no wear track could be seen on the diamond disc surfaces sliding against silicon nitride balls—as also it was observed against alumina balls in authors' previous paper [131]. Only the balls worn out and the presented wear data of balls are in tandem with the reported literatures.

Moreover, the lowest COF of ~0.03 obtained using machine oil lubricant in the present paper appears to be better than liquid bio-diesel or solid grapheme lubricant used for diamond coatings previously.

4. Conclusions

As-grown, annealed, polished CVD grown diamonds have been characterised adopting some new techniques which are conventionally used for powder bulk specimens. Use of TGA-DSC analysis could pinpoint the diamond oxidation temperature for the first time. Attempt has been made to quantify the porosity that might be present in the freestanding coating, which reduces the theoretical density values of CVD grown diamond. It has been found that the internal defects that are present inside diamond lattice determine the black or white nature of diamond coatings. Qualitative analysis of such defects that are present in different grades of PCDs (line, plane or volume defects) like dislocations, grain boundary, twining has been revealed by transmission electron microscope. Whereas, the other atomic defects, like nitrogen vacancy, atomic hydrogen, are revealed by FTIR and PL studies. Raman signals quantify the internal stress, crystallinity and the quality of the CVD grown diamond coatings. It has been found that after polishing the surface of diamond becomes very unclean due to contamination from the polishing consumables and generation of non-diamond carbon phases. Use of machine oil lubricant for tribological application of such polished diamond surfaces has found to be very effective in reducing the COF values considerably. However, the CVD grown diamond nucleation side produced best COF value of about 0.03 with 3.95×10^{-5} $mm^3/N\ m$ wear rate for the silicon nitride counterface.

Acknowledgements

AKM is presently on Extraordinary Leave (EOL) from CSIR-CGCRI and working as FWO postdoctoral researcher at IMO-IMOMEC, University of Hasselt, Belgium. The work presented here was carried out during 1 year training programme on different instruments at the CSIR-CGCRI by the author. He acknowledges help in taking measurements and preparing the training report from research fellows and institute colleagues during his training programme, like Mr. Aniruddha Samanta, Ms. Debarati Ghosh, Ms. Snigdha Roy, to name a few, of the different divisions of CSIR-CGCRI.

Author details

Awadesh Kumar Mallik[1,2,3]

1 CSIR—Central Glass and Ceramic Research Institute, Kolkata, West Bengal, India

2 Institute for Materials Research (IMO), Hasselt University, Diepenbeek, Belgium

3 IMOMEC, IMEC vzw, Diepenbeek, Belgium

*Address all correspondence to: amallik@cgcri.res.in

References

[1] Gicquel A, Hassouni K, Silva F, Achard J. CVD diamond films: From growth to applications. Current Applied Physics. 2001;**1**:479-496. DOI: 10.1016/S1567-1739(01)00061-X

[2] Balmer RS, Brandon JR, Clewes SL, Dhillon HK, Dodson JM, Friel I, et al. Chemical vapour deposition synthetic diamond: Materials, technology and applications. Journal of Physics: Condensed Matter. 2009;**21**:364221. DOI: 10.1088/0953-8984/21/36/364221

[3] Nemanich RJ, Carlisle JA, Hirata A, Haenen K, guest editors. CVD diamond—Research, applications, and challenges. MRS Bulletin. 2014;**39**:490-494. DOI: 10.1557/mrs.2014.97

[4] Spear KE. Diamond-ceramic coating of the future. Journal of the American Ceramic Society. 1989;**72**:171-191

[5] May PW. Diamond thin films: A 21st-century material. Philosophical Transactions of the Royal Society of London. 2000;**358**:473-495

[6] Eaton-Magaña S, Shigley JE. Observations on CVD-grown synthetic diamonds: A review. Gems & Gemology. 2016;**52**:222-245

[7] Chernov VV, Gorbachev AM, Vikharev AL, Lobaev MA. Criterion for comparison of MPACVD reactors working at different microwave frequencies and diamond growth conditions. Physica Status Solidi A. 2016;**213**:2564-2569. DOI: 10.1002/pssa.201600193

[8] Teii K. Plasma deposition of diamond at low pressures: A review. IEEE Transactions on Plasma Science. 2014;**42**:3862-3869. DOI: 10.1109/TPS.2014.2333772

[9] Kasu M. Diamond epitaxy: Basics and applications. Progress in Crystal Growth and Characterization of Materials. 2016;**62**:317-328. DOI: 10.1016/j.pcrysgrow.2016.04.017

[10] Linnik SA, Gaydaychuk AV. Synthesis of multilayer polycrystalline diamond films using bias-induced secondary nucleation. Materials Letters. 2015;**139**:389-392. DOI: 10.1016/j.matlet.2014.10.142

[11] Schwander M, Partes K. A review of diamond synthesis by CVD processes. Diamond & Related Materials. 2011;**20**:1287-1301. DOI: 10.1016/j.diamond.2011.08.005

[12] Kromka A, Babchenko O, Izak T, Hruska K, Rezek B. Linear antenna microwave plasma CVD deposition of diamond films over large areas. Vacuum. 2012;**86**:776-779. DOI: 10.1016/j.vacuum.2011.07.008

[13] Nor RM, Bakar SA, Thandavan TM, Rusop M. Diamond: Synthesis, characterisation and applications, carbon and oxide nanostructures. Advanced Structured Materials. 2010;**5**:195-217

[14] Muchnikov AB, Vikharev AL, Radishev DB, Isaev VA, Ivanov OA, Gorbachev AM. A wafer of combined single-crystalline and polycrystalline CVD diamond. Materials Letters. 2015;**139**:1-3. DOI: 10.1016/j.matlet.2014.10.022

[15] Kar R, Patel NN, Chand N, Shilpa RK, Dusane RO, Patil DS, et al. Detailed investigation on the mechanism of co-deposition of different carbon nanostructures by microwave plasma CVD. Carbon. 2016;**106**:233-242. DOI: 10.1016/j.carbon.2016.05.027

[16] Goncalves JAN, Sandonato GM, Iha K. Characterization of boron doped CVD diamond films by Raman spectroscopy and X-ray diffractometry.

Diamond and Related Materials. 2002;**11**:1578-1583. DOI: 10.1016/S0925-9635(02)00103-6

[17] Fries MD, Vohra YK. Properties of nanocrystalline diamond thin films grown by MPCVD for biomedical implant purposes. Diamond and Related Materials. 2004;**13**:1740-1743. DOI: 10.1016/j.diamond.2004.02.014

[18] Palyanov YN, Kupriyanov IN, Khokhryakov AF, Ralchenko VG. Crystal Growth of Diamond, Handbook of Crystal Growth. Bulk Crystal Growth, A Volume in Handbook of Crystal Growth. 2nd ed. 2015. pp. 671-713. DOI: 10.1016/B978-0-444-63303-3.00017-1

[19] Terranova ML, Rossi M, Tamburri E. Nanocrystalline sp2 and sp3 carbons: CVD synthesis and applications. Crystallography Reports. 2016;**61**:897-906. DOI: 10.1134/S1063774516060158

[20] Bogatskiy A, Butler JE. A geometric model of growth for cubic crystals: Diamond. Diamond & Related Materials. 2015;**53**:58-65. DOI: 10.1016/j.diamond.2014.12.010

[21] Hua C, Yan X, Wei J, Guo J, Liu J, Chen L, et al. Intrinsic stress evolution during different growth stages of diamond film. Diamond & Related Materials. 2017;**73**:62-66. DOI: 10.1016/j.diamond.2016.12.008

[22] Lobanov SS, Prakapenka VB, Prescher C, Konopkov Z, Liermann HP, Crispin KL, et al. Pressure, stress, and strain distribution in the double-stage diamond anvil cell. Journal of Applied Physics. 2015;**118**:035905. DOI: 10.1063/1.4927213

[23] Nedosekin P, Gladchenkov E, Zakharchenko K, Kolyubin V. Review the space radiation CVD diamond multi-layer detector. Moscow: International Siberian Conference on Control and Communications (SIBCON). 2016. pp. 1-4. DOI: 10.1109/SIBCON.2016.7491780

[24] Cui Y, Wang W, Shen B, Sun F. Reprint of "A study of CVD diamond deposition on cemented carbide ball-end milling tools with high cobalt content using amorphous ceramic interlayers". Diamond & Related Materials. 2016;**63**:51-59. DOI: 10.1016/j.diamond.2016.01.017

[25] Shen B, Sun FH, Zhang ZM, Shen HS, Guo SS. Application of ultra-smooth composite diamond film coated WC–Co drawing dies under water-lubricating conditions. Transactions of Nonferrous Metals Society of China. 2013;**23**:161-169. DOI: 10.1016/S1003-6326(13)62443-7

[26] Shen B, Sun F. Deposition and friction properties of ultra-smooth composite diamond films on Co-cemented tungsten carbide substrates. Diamond and Related Materials. 2009;**18**:238-243. DOI: 10.1016/j.diamond.2008.10.053

[27] Dumpala R, Chandran M, Ramachandra Rao MS. Engineered CVD diamond coatings for machining and tribological applications. JOM. 2015;**67**:1565-1577. DOI: 10.1007/s11837-015-1428-2

[28] Tian Q, Yang B, Zhuang H, Guo Y, Wang C, Zhai Z, et al. Hybrid diamond/graphite films: Morphological evolution, microstructure and tribological properties. Diamond & Related Materials. 2016;**70**:179-185. DOI: 10.1016/j.diamond.2016.10.020

[29] Nagasaka H, Ito K, Mori J, Shimizu T, Sasaki S. Tribological properties of polycrystalline diamond films prepared by hot-filament chemical vapor deposition methods. In: Proceedings of the 16th International Conference on

Nanotechnology; 22-25 August 2016. Sendai, Japan; 2016. pp. 616-619

[30] Shen B, Chen S, Sun F. Investigation on the long-duration tribological performance of bilayered diamond/ diamond-like carbon films. Proceedings of the Institution of Mechanical Engineers, Part J: Journal of Engineering Tribology. 2014;**228**:628-641. DOI: 10.1177/1350650114522611

[31] Zhang D, Shen B, Sun F. Study on tribological behavior and cutting performance of CVD diamond and DLC films on Co-cemented tungsten carbide substrates. Applied Surface Science. 2010;**256**:2479-2489. DOI: 10.1016/j. apsusc.2009.10.092

[32] Xinchang W, Liang W, Bin S, Fanghong S. Friction and wear performance of boron doped, Undoped microcrystalline and fine grained composite diamond films. Chinese Journal of Mechanical Engineering. 2015;**28**:155-163. DOI: 10.3901/ CJME.2014.1114.168

[33] Chen N, Pu L, Sun F, He P, Zhu Q, Ren J. Tribological behavior of HFCVD multilayer diamond film on silicon carbide. Surface & Coatings Technology. 2015;**272**:66-71. DOI: 10.1016/j.surfcoat.2015.04.023

[34] Jones AN, Ahmed W, Rego CA, Taylor H, Beake BD, Jackson MJ. Investigation of the tribological properties of diamond films. Journal of Materials Engineering and Performance. 2007;**16**:131-134. DOI: 10.1007/ s11665-006-9021-z

[35] Kumaran CR, Tiwari B, Chandran M, Bhattacharya SS, Ramachandra Rao MS. Effect of temperature on the stability of diamond particles and continuous thin films by Raman imaging. Journal of Nanoparticle Research. 2013;**15**:1509

[36] Ren B, Huang J, Yu H, Yang W, Wang L, Pan Z, et al. Thermal stability of hydrogenated diamond films in nitrogen ambience studied by reflection electron energy spectroscopy and X-ray photoelectron spectroscopy. Applied Surface Science. 2016;**388**:565-570. DOI: 10.1016/j.apsusc.2015.10.067

[37] Nakazawa H, Okuno S, Magara K, Nakamura K, Miura S, Enta Y. Tribological properties and thermal stability of hydrogenated, silicon/ nitrogen-coincorporated diamond-like carbon films prepared by plasma-enhanced chemical vapor deposition. Japanese Journal of Applied Physics. 2016;**55**:12550. DOI: 10.7567/ JJAP.55.125501

[38] Zolotukhin A, Kopylov PG, Ismagilov RR, Obraztsov AN. Thermal oxidation of CVD diamond. Diamond & Related Materials. 2010;**19**:1007-1011. DOI: 10.1016/j.diamond.2010.03.005

[39] Fecher J, Wormser M, Rosiwal SM. Long term oxidation behavior of micro- and nano-crystalline CVD diamond foils. Diamond & Related Materials. 2016;**61**:41-45. DOI: 10.1016/j. diamond.2015.11.009

[40] Lee J, Anderson MW, Gray FA, John P, Lee J, Baik Y, et al. Oxidation of CVD diamond powders. Diamond and Related Materials. 2004;**13**:1070-1074. DOI: 10.1016/j.diamond.2004.01.026

[41] Joung YH, Kang FS, Lee S, Kang H, Choi WS, Choi YK, et al. Reaction gas ratio effect on the growth of a diamond film using microwave plasma-enhanced chemical vapor deposition. Journal of Nanoscience and Nanotechnology. 2016;**16**:5295-5297. DOI: 10.1166/ jnn.2016.12212

[42] Lloret F, Araujo D, Eon D, Villar MP, Gonzalez-Leal JM, Bustarret E. Influence of methane concentration on MPCVD overgrowth of 100-oriented etched diamond substrates. Physica Status Solidi A. 2016;**213**:2570-2574. DOI: 10.1002/pssa.201600182

[43] Izak T, Davydova M, Varga M, Potocky S, Kromka A. Growth rate enhancement and morphology engineering of diamond films by adding CO_2 or N_2 in hydrogen rich gas chemistry. Advanced Science, Engineering and Medicine. 2014;**6**:749-755. DOI: 10.1166/asem.2014.1571

[44] Yang L, Jiang C, Guo S, Zhang L, Gao J, Peng J, et al. Novel diamond films synthesis strategy: Methanol and argon atmosphere by microwave plasma CVD method without hydrogen. Nanoscale Research Letters. 2016;**11**:415. DOI: 10.1186/s11671-016-1628-x

[45] Liu C, Wang JH, Weng J. Growth of micro-and nanocrystalline dual layer composite diamond films by microwave plasma CVD: Influence of CO_2 concentration on growth of nano-layer. Journal of Crystal Growth. 2015;**410**:30-34. DOI: 10.1016/j.jcrysgro.2014.10.040

[46] Paxton WF, Howell M, Kang WP, Davidson JL. Influence of hydrogen on the thermionic electron emission from nitrogen-incorporated polycrystalline diamond films. Journal of Vacuum Science & Technology B. 2012;**30**:021202. DOI: 10.1116/1.3684982

[47] Zou Y, Larsson K. Effect of boron doping on the CVD growth rate of diamond. Journal of Physical Chemistry C. 2016;**120**:10658-10666. DOI: 10.1021/acs.jpcc.6b02227

[48] Lobaev MA, Gorbachev AM, Bogdanov SA, Vikharev AL, Radishev DB, Isaev VA, et al. Influence of CVD diamond growth conditions on nitrogen incorporation. Diamond & Related Materials. 2017;**72**:1-6. DOI: 10.1016/j.diamond.2016.12.011

[49] Zaitsev AM, Wang W, Moe KS, Johnson P. Spectroscopic studies of yellow nitrogen-doped CVD diamonds. Diamond & Related Materials. 2016;**68**:51-61. DOI: 10.1016/j.diamond.2016.06.002

[50] Ficek M, Sankaran KJ, Ryl J, Bogdanowicz R, Lin I, Haenen K, et al. Ellipsometric investigation of nitrogen doped diamond thin films grown in microwave $CH_4/H_2/N_2$ plasma enhanced chemical vapor deposition. Applied Physics Letters. 2016;**108**:241906. DOI: 10.1063/1.4953779

[51] Kono S, Teraji T, Kodama H, Sawabe A. Reprint of "imaging of diamond defect sites by electron-beam-induced current". Diamond & Related Materials. 2016;**63**:30-37. DOI: 10.1016/j.diamond.2016.01.020

[52] Khan RUA, Cann BL, Martineau PM, Samartseva J, Freeth JJP, Sibley SJ, et al. Colour-causing defects and their related optoelectronic transitions in single crystal CVD diamond. Journal of Physics: Condensed Matter. 2013;**25**:275801. DOI: 10.1088/0953-8984/25/27/275801

[53] Prokhorova IA, Voloshinb AE, Ralchenko VG, Bolshakov AP, Romanove DA, Khomich AA, et al. X-ray diffraction characterization of epitaxial CVD diamond films with natural and isotopically modified compositions. Crystallography Reports. 2016;**61**:979-986. DOI: 10.1134/S1063774516060122

[54] Chowdhury S, Laugier MT, Henry J. XRD stress analysis of CVD diamond coatings on SiC substrates. International Journal of Refractory Metals & Hard Materials. 2007;**25**:39-45. DOI: 10.1016/j.ijrmhm.2005.11.012

[55] Willems B, Tallaire A, Achard J. Optical study of defects in thick undoped CVD synthetic diamond layers. Diamond & Related Materials. 2014;**41**:25-33. DOI: 10.1016/j.diamond.2013.09.010

[56] Collins AT. The characterisation of point defects in diamond by luminescence spectroscopy. Diamond

and Related Materials. 1992;**1**:457-469. DOI: 10.1016/0925-9635(92)90146-F

[57] Zaitsev AM, Moe KS, Wang W. Optical centers and their depth distribution in electron irradiated CVD diamond. Diamond & Related Materials. 2017;**71**:38-52. DOI: 10.1016/j.diamond.2016.11.015

[58] Ralchenko V, Sedov V, Saraykin V, Bolshakov A, Zavedeev E, Ashkinazi E, et al. Precise control of photoluminescence of silicon-vacancy color centers in homoepitaxial single-crystal diamond: Evaluation of efficiency of Si doping from gas phase. Applied Physics A: Materials Science & Processing. 2016;**122**:795. DOI: 10.1007/s00339-016-0343-x

[59] Walker J. Optical absorption and luminescence in diamond. Reports on Progress in Physics. 1979;**42**:1606-1659. DOI: 10.1088/0034-4885/42/10/001

[60] Sawada H, Ichinose H. Atomic structure of fivefold twin center in diamond film. Diamond & Related Materials. 2005;**14**:109-112. DOI: 10.1016/j.diamond.2004.07.016

[61] Purwanto S, Iskandar R, Dimyati A. Scanning transmission electron microscopy (STEM) study on surface modified CVD diamond/Si(111) film post implanted Fe-B and NiFe-B related to GMR properties. AIP Conference Proceedings. 2016;**1725**:020064. DOI: 10.1063/1.4945518

[62] Ma GHM, Lee YH, Glass JT. Electron microscopic characterization of diamond films grown on Si by bias-controlled chemical vapor deposition. Journal of Materials Research. 1990;**5**:2367-2377. DOI: 10.1557/JMR.1990.2367

[63] Quin LC, Zhou D, Krauss AR, Gruen DM. TEM characterisation of nanodiamond thin films. Nanostructured Materials. 1998;**10**:649-660. DOI: 10.1016/S0965-9773(98)00092-0

[64] Nistor LC, Landuyt JV, Ralchenko VG, Obraztsova ED, Smolin AA. Nanocrystalline diamond films: Transmission electron microscopy and Raman spectroscopy characterisation. Diamond and Related Materials. 1997;**6**:159-168. DOI: 10.1016/S0925-9635(96)00743-1

[65] Pal KS, Bysakh S, Mallik AK, Dandapat N, Datta S, Guha BK. Influence of growth conditions on microstructure and defects in diamond coatings grown by microwave plasma enhanced CVD. Bulletin of Materials Science. 2015;**38**:1-8. Available from: http://www.ias.ac.in/article/fulltext/boms/038/03/0717-0724

[66] Berger D, Uhlmann E, Dethlefs I. A novel method for specimen preparation and analysis of CVD diamond coated tools using focussed ion beams (FIB) and scanning electron microscopy (SEM). Metallography, Microstructure, and Analysis. 2015;**4**:49-57. DOI: 10.1007/s13632-014-0184-y

[67] Sawada H, Ichinose H, Kohyama M. Atomic structure of the R3 and R9 grain boundaries in CVD diamond film. Scripta Materialia. 2004;**51**:689-692. DOI: 10.1016/j.scriptamat.2004.06.011

[68] Porro S, Temmerman GD, MacLaren DA, Lisgo S, Rudakov DL, Westerhout J, et al. Surface analysis of CVD diamond exposed to fusion plasma. Diamond & Related Materials. 2010;**19**:818-823. DOI: 10.1016/j.diamond.2010.01.051

[69] Zhang M, Xia Y, Wang L, Gu B, Su Q, Lou Y. Effects of the deposition conditions and annealing process on the electric properties of hot-filament CVD diamond films. Journal of Crystal Growth. 2005;**274**:21-27. DOI: 10.1016/j.jcrysgro.2004.09.092

[70] Yan X, Wei J, Guo J, Hua C, Liu J, Chen L, et al. Mechanism of graphitization and optical degradation of CVD diamond films by rapid heating treatment. Diamond and Related Materials. 2017;**73**:39-46. DOI: 10.1016/j.diamond.2016.11.010

[71] Khomich AA, Kudryavtsev OS, Dolenko TA, Shiryaev AA, Fisenko AV, Konov VI, et al. Anomalous enhancement of nanodiamond luminescence upon heating. Laser Physics Letters. 2017;**14**:025702. DOI: 10.1088/1612-202X/aa52f5

[72] Khomich A, Ralchenko VG, Nistor L, Vlasov I, Khmelnitskiy R. Optical properties and defect structure of CVD diamond films annealed at 900-1600°C. Physica Status Solidi (a). 2000;**181**:37-44. DOI: 10.1002/1521-396X(200009)181:1<37::AID-PSSA37>3.0.CO;2-T

[73] Yamamoto T, Umeda T, Watanabe K, Onoda S, Markham ML, Twitchen DJ, et al. Extending spin coherence times of diamond qubits by high temperature annealing. Physical Review B. 2013;**88**:075206. DOI: https://doi.org/10.1103/PhysRevB.88.075206

[74] Khomich AV, Ralchenko VG, Vlasov AV, Khmelnitskiy RA, Vlasov II, Konov VI. Effect of high temperature annealing on optical and thermal properties of CVD diamond. Diamond and Related Materials. 2001;**10**:546-551. DOI: 10.1016/S0925-9635(00)00517-3

[75] Ralchenko VG, Nistor L, Pleuler E, Khomich A, Vlasov I, Khmelnitskiy R. Structure and properties of high-temperature annealed CVD diamond. Diamond and Related Materials. 2003;**12**:1964-1970. DOI: 10.1016/S0925-9635(03)00214-0

[76] Seshan V, Ullien D, Castellanos-Gomez A, Sachdeva S, Murthy DHK, Savenije TJ, et al. Hydrogen termination of CVD diamond films by high-temperature annealing at atmospheric pressure. The Journal of Chemical Physics. 2013;**138**:234707. DOI: 10.1063/1.4810866

[77] Dychalska A, Fabisiak K, Paprocki K, Makowiecki J, Iskaliyeva A, Szybowicz M. A Raman spectroscopy study of the effect of thermal treatment on structural and photoluminescence properties of CVD diamond films. Materials and Design. 2016;**112**:320-327. DOI: 10.1016/j.matdes.2016.09.092

[78] Sanzenbacher LM. Raman Spectroscopic Studies of Single Crystal Diamond [MS Thesis]. Graduate Program in Chemical Physics The Ohio State University; 2011. pp. 1-41

[79] Ferrari AC, Robertson J. Raman spectroscopy of amorphous, nanostructured, diamond-like carbon, and nanodiamond. Philosophical Transactions of the Royal Society of London. Series A. 2004;**362**:2477-2512. DOI: 10.1098/rsta.2004.1452

[80] Haubnera R, Rudigiera M. Raman characterisation of diamond coatings using different laser wavelengths. Physics Procedia. 2013;**46**:71-78. DOI: 10.1016/j.phpro.2013.07.047

[81] Prawer S, Nemanich RJ. Raman spectroscopy of diamond and doped diamond. Philosophical Transactions of the Royal Society of London. Series A. 2004;**362**:2537-2565. DOI: 10.1098/rsta.2004.1451

[82] Knight DS, White WB. Characterization of diamond films by Raman spectroscopy. Journal of Materials Research. 1989;**4**:385-393. DOI: 10.1557/JMR.1989.0385

[83] Pandey M, D'Cunha R, Tyagi AK. Defects in CVD diamond: Raman and XRD studies. Journal of Alloys and Compounds. 2002;**333**:260-265. DOI: 10.1016/S0925-8388(01)01740-6

[84] Bachmann PK, Bausen HD, Lade H, Leers D, Wiechert DU, Herres N, et al. Raman and X-ray studies of polycrystalline CVD diamond films. Diamond and Related Materials. 1994;**3**:1308-1314. DOI: 10.1016/0925-9635(94)901430

[85] Solin SA, Ramdas AK. Raman spectrum of diamond. Physical Review B. 1970;**1**:1687, 1698. DOI: 10.1103/PhysRevB.1.1687

[86] Filik J. Raman spectroscopy: A simple, non-destructive way to characterize diamond and diamond-like carbon materials. Spectroscopy Europe. 2005;**17**:1010-1017

[87] Chen N, Ai J, Chen Y, He P, Ren J, Ji D. Multilayer strategy and mechanical grinding for smoothing CVD diamond coated defective substrate. Materials and Design. 2016;**103**:194-200. DOI: 10.1016/j.matdes.2016.04.069

[88] Kellermann K, Ehrhardt S, Fandrey J, Rosiwal SM, Singer RF. Influence of surface roughness on the tribological properties of HF-CVD diamond coated heat-treatable steel. Wear. 2010;**269**:811-815. DOI: 10.1016/j.wear.2010.08.009

[89] Schade A, Rosiwal SM, Singer RF. Influence of surface topography of HF-CVD diamond films on self-mated planar sliding contacts in dry environments. Surface & Coatings Technology. 2007;**201**:6197-6205. DOI: 10.1016/j.surfcoat.2006.11.024

[90] Zhang R, Lub Z, Shi W, Leng S, Tang B. Low friction of diamond sliding against Al2O3 ceramic ball based on the first principles calculations. Surface & Coatings Technology. 2015;**283**:129-134. DOI: 10.1016/j.surfcoat.2015.10.062

[91] Wang X, Shen X, Zhao T, Sun F, Shen B. Tribological properties of SiC-based MCD films synthesized using different carbon sources when sliding against Si_3N_4. Applied Surface Science. 2016;**369**:448-459. DOI: 10.1016/j.apsusc.2016.01.249

[92] Miyake S, Shindo T, Miyake M. Friction properties of surface-modified polished chemical-vapor-deposited diamond films under boundary lubrication with water and poly-alpha olefin. Tribology International. 2016;**102**:287-296. DOI: 10.1016/j.triboint.2016.05.026

[93] Fazio LD, Syngellakis S, Wood RJK, Fugiuele FM, Sciume G. Nanoindentation of CVD diamond: Comparison of an FE model with analytical and experimental data. Diamond and Related Materials. 2001;**10**:765-769. DOI: 10.1016/S0925-9635(00)00496-9

[94] Nitti MA, Cicala G, Brescia R, Romeo A, Guion JB, Perna G, et al. Mechanical properties of MWPECVD diamond coatings on Si substrate via nanoindentation. Diamond & Related Materials. 2011;**20**:221-226. DOI: 10.1016/j.diamond.2010.12.002

[95] Chowdhury S, de Barra E, Laugier MT. Hardness measurement of CVD diamond coatings on SiC substrates. Surface Coatings & Technology. 2005;**193**:200-205. DOI: 10.1016/j.surfcoat.2004.08.131

[96] Mohr M, Picollo F, Battiato A, Bernardi E, Forneris J, Tengattini A, et al. Characterization of the recovery of mechanical properties of ion-implanted diamond after thermal annealing. Diamond & Related Materials. 2015;**63**:75-79. DOI: 10.1016/j.diamond.2015.11.008

[97] Wiora M, Bruehne K, Floeter A, Gluche P, Willey TM, Kucheyev SO, et al. Grain size dependent mechanical properties of nanocrystalline diamond films grown by hot-filament CVD. Diamond & Related Materials.

2009;**18**:927-930. DOI: 10.1016/j.diamond.2008.11.026

[98] Cicala G, Magaletti V, Senesi GS, Carbone G, Altamura D, Giannini C, et al. Superior hardness and Young's modulus of low temperature nanocrystalline diamond coatings. Materials Chemistry and Physics. 2014;**144**:505-511. DOI: 10.1016/j.matchemphys.2014.01.027

[99] Hei LF, Lu FX, Li CM, Tang WZ, Chen GC, Song JH. A review on mechanical properties of freestanding diamond films. Advanced Materials Research. 2012;**490-495**:3059-3064. DOI: 10.4028/www.scientific.net/AMR.490-495.3059

[100] Kamiya S, Takahashi H, Kobayashi A, Saka M, Abe H. Fracture strength of chemically vapor deposited diamond on the substrate and its relation to the crystalline structure. Diamond and Related Materials. 2000;**9**:1110-1114. DOI: 10.1016/S0925-9635(99)00331-3

[101] Kulisch W, Popov C, Boychev S, Buforn L, Favaro G, Conte N. Mechanical properties of nano-crystalline diamond/amorphous carbon composite films prepared by microwave plasma chemical vapour deposition. Diamond & Related Materials. 2004;**13**:1997-2002. DOI: 10.1016/j.diamond.2004.04.002

[102] Wang X, Shen X, Sun F, Shen B. Mechanical properties and solid particle erosion of MCD films synthesized using different carbon sources by BE-HFCVD. International Journal of Refractory Metals and Hard Materials. 2016;**54**:370-377. DOI: 10.1016/j.ijrmhm.2015.09.004

[103] Chu PK, Li L. Characterization of amorphous and nanocrystalline carbon films. Materials Chemistry and Physics. 2006;**96**:253-277. DOI: 10.1016/j.matchemphys.2005.07.048

[104] Pradhan D, Lin IN. Grain-size-dependent diamond-nondiamond composite films: Characterization and field-emission properties. ACS Applied Materials & Interfaces. 2009;**1**:1444-1450. DOI: 10.1021/am9001327

[105] Yanchuk IB, Valakh MY, Vul AY, Golubev VG, Grudinkin SA, Feoktistov NA, et al. Raman scattering, AFM and nanoindentation characterisation of diamond films obtained by hot filament CVD. Diamond and Related Materials. 2004;**13**:266-269. DOI: 10.1016/j.diamond.2003.11.001

[106] Sharda T, Rahaman MM, Nukaya Y, Soga T, Jimbo T, Umeno M. Structural and optical properties of diamond and nano-diamond films grown by microwave plasma chemical vapor deposition. Diamond and Related Materials. 2001;**10**:561-567. DOI: 10.1016/S0925-9635(00)00390-3

[107] Yin Z, Akkerman Z, Yang BX, Smith FW. Optical properties and microstructure of CVD diamond films. Diamond and Related Materials. 1997;**6**:153-158. DOI: 10.1016/S0925-9635(96)00740-6

[108] Yang JX, Li CM, Lu FX, Chen GC, Tang WZ, Tong YM. Microstructure and fracture strength of different grades of freestanding diamond films deposited by a DC arc plasma jet process. Surface & Coatings Technology. 2005;**192**:171-176. DOI: 10.1016/j.surfcoat.2004.04.089

[109] Yang JX, Duan XF, Lu FX, Li CM, Zuo TC, Wang FL. The influence of dark feature on optical and thermal property of DC arc plasma jet CVD diamond films. Diamond & Related Materials. 2005;**14**:1583-1587. DOI: 10.1016/j.diamond.2005.03.010

[110] Calvani P, Bellucci A, Girolami M, Orlando S, Valentini V, Polini R, et al. Black diamond for solar energy

conversion. Carbon. 2016;**105**:401-407. DOI: 10.1016/j.carbon.2016.04.017

[111] Bangert U, Barnes R. Electron energy loss spectroscopy of defects in diamond. Physica Status Solidi (a). 2007;**204**:2201-2210. DOI: 10.1002/pssa.200675442

[112] Badzian A. The displacement disorder of atoms in diamond crystals revealed by X-ray imaging plate detector. Diamond & Related Materials. 2016;**69**:19-32. DOI: 10.1016/j.diamond.2016.07.004

[113] Zolotukhin AA, Ismagilov RR, Dolganov MA, Obraztsov AN. Morphology and Raman spectra peculiarities of chemical vapor deposition diamond films. Journal of Nanoelectronics and Optoelectronics. 2012;**7**:22-28. DOI: 10.1166/jno.2012.1210

[114] Miki H, Tsutsui A, Takeno T, Takagi T. Friction properties of partially polished CVD diamond films at different sliding speeds. Diamond & Related Materials. 2012;**24**:167-170. DOI: 10.1016/j.diamond.2012.01.004

[115] Mallik AK, Bysakh S, Pal KS, Dandapat N, Guha BK, Datta S, et al. Large area deposition of polycrystalline diamond coatings by microwave plasma CVD. Transactions of the Indian Ceramic Society. 2013;**72**:225-232. DOI: 10.1080/0371750X.2013.870768

[116] Mallik AK, Bysakh S, Pal KS, Dandapat N, Guha BK, Datta S, et al. Synthesis and characterisation of freestanding diamond coatings. Indian Journal of Engineering and Materials Science. 2013;**20**:522-532

[117] Mallik AK, Bhar R, Bysakh S. An effort in planarising microwave plasma CVD grown polycrystalline diamond (PCD) coated Si wafers of 4 inch diameter. Materials Science in Semiconductor Processing. 2016;**43**:1-7. DOI: 10.1016/j.mssp.2015.11.016

[118] Mallik AK, Mendes JC, Rotter SZ, Bysakh S. Detonation nanodiamond seeding technique for nucleation enhancement of CVD diamond—Some experimental insights. Advances in Ceramic Science and Engineering. 2014;**3**:36-45. DOI: 10.14355/acse.2014.03.005

[119] Mallik AK, Bysakh S, Bhar R, Rotter SZ, Mendes JC. Effect of seed size, suspension recycling and substrate pre-treatment on the CVD growth of diamond coatings. Open Journal of Applied Sciences. 2015;**5**:747-763. DOI: 10.4236/ojapps.2015.512071

[120] Determination of the Specific Surface Area of Solids by Gas Adsorption—BET Method. 2nd ed. BS ISO 9277:2010

[121] Pal KS, Mallik AK, Dandapat N, Ray NR, Datta S, Bysakh S, et al. Microscopic properties of MPCVD diamond coatings studied by micro-Raman and micro-photoluminescence spectroscopy. Bulletin of Materials Science. 2015;**38**:537-549. DOI: 10.1007/s12034-015-0860-9

[122] Tallaire A, Collins AT, Charles D, Achard J, Sussmann R, Gicquel A, et al. Characterisation of high-quality thick single-crystal diamond grown by CVD with a low nitrogen addition. Diamond & Related Materials. 2006;**15**:1700-1707. DOI: 10.1016/j.diamond.2006.02.005

[123] Available from: https://www.bruker.com/fileadmin/user_upload/8-PDF-Docs/OpticalSpectrospcopy/FT-IR/ALPHA/AN/AN81_Diamonds_EN.pdf [Accessed: 12 April 2017]

[124] Tang CJ, Hou H, Fernandes AJS, Jiang XF, Pinto JL, Ye H. Investigation of bonded hydrogen defects in nanocrystalline diamond films grown with nitrogen/methane/hydrogen plasma at high power conditions. Journal of Crystal Growth. 2017;**460**:16-22. DOI: 10.1016/j.jcrysgro.2016.12.050

[125] Pezzagna S, Rogalla D, Wildanger D, Meijer J, Zaitsev A. Creation and nature of optical centres in diamond for single-photon emission—Overview and critical remarks. New Journal of Physics. 2011;**13**:035024. DOI: 10.1088/1367-2630/13/3/035024

[126] Schirhagl R, Chang K, Loretz M, Degen CL. Nitrogen-vacancy centers in diamond: Nanoscale sensors for physics and biology. Annual Review of Physical Chemistry. 2014;**65**:83-105. DOI: 10.1146/annurev-physchem-040513-103659

[127] Ulrika FS, D'Haenens-Johansson AK, Moe KS, Johnson P, Wang W. Large colorless HPHT-grown synthetic gem diamonds from new diamond technology. Russia, Gems & Gemology. 2015;**51**:3. Available from: https://www.gia.edu/gems-gemology/fall-2015-large-colorless-hpht-grown-synthetic-gem-diamond-technology-russiaFacebook48TwitterPrintEmailPinterestMore97

[128] Collins AT. Vacancy enhanced aggregation of nitrogen in diamond. Journal of Physics C: Solid State Physics. 1980;**13**:2641-2650. DOI: 10.1088/0022-3719/13/14/006

[129] Mallik AK, Bysakh S, Sreemany M, Roy S, Ghosh J, Roy S, et al. Property mapping of polycrystalline diamond coatings over large area. Journal of Advanced Ceramics. 2014;**3**:56-70. DOI: 10.1007/s40145-014-0093-1

[130] Oliver WC, Pharr GM. An improved technique for determining hardness and elastic modulus. Journal of Materials Research. 1992;7:1564-1583. DOI: 10.1557/JMR.1992.1564

[131] Jana A, Dandapat N, Das M, Balla VK, Chakraborty S, Saha R, et al. Severe wear behaviour of alumina balls sliding against diamond ceramic coatings. Bulletin of Materials Science. 2016;**39**:573-586. DOI: 10.1007/s12034-016-1166-2

[132] Almeida FA, Maru MM, Shabani M, Oliveira FJ, Silva RF, Achete CA. Enhancing the tribological performance under biodiesel lubrication using CVD diamond coated parts. Wear. 2013;**302**:1370-1377. DOI: 10.1016/j.wear.2013.01.090

[133] Shen B, Chen S, Chen Y, Sun F. Enhancement on the tribological performance of diamond films by utilizing graphene coating as a solid lubricant. Surface & Coatings Technology. 2017;**311**:35-45. DOI: 10.1016/j.surfcoat.2016.12.094

[134] Chen S, Shen B, Sun F. Tribo-map of CVD diamond film sliding against silicon nitride in air. Key Engineering Materials. 2014;**589-590**:405-410. DOI: 10.4028/www.scientific.net/KEM.589-590.405

[135] Bogatov A, Traksmaa R, Podgursky V. Changes in surface morphology, deflection and wear of microcrystalline diamond film observed during sliding tests against Si_3N_4 balls. Key Engineering Materials. 2016;**674**:145-115. DOI: 10.4028/www.scientific.net/KEM.674.145

Chapter 3

Significance of Diamond as a Cutting Tool in Ultra-Precision Machining Process

P. Suya Prem Anand

Abstract

This chapter focuses on the purpose of using diamond as a cutting tool in various ultra-precision machining applications. The complicated structures such as resin and ceramic mold used for making optical lenses are machined by the diamond tool to improve the precision of the finished product. It is difficult to machine hard and brittle materials such as glasses, ceramics, and composites with the assistance of diamond tool due to the complexity in the aspheric surfaces. Moreover, the tool wear is a major problem in machining these hard materials to a fine dimensional accuracy and tolerances. The microscopic defect forms at the cutting edge lead to the damage of the surface finish of the workpiece material. Therefore, the discussions are associated with the achievement of machining hard materials using a diamond tool in ultra-precision applications.

Keywords: cutting tool, optical lenses, ceramics, glasses, diamond coating, tool wear, AE sensor

1. Introduction

The ultra-precision machining is the process of improving the dimensional accuracy and the surface quality of the product in the submicron level. Therefore, the magnitude of surface roughness must be less than 10 nm, and the machining accuracy will be in the range of 0.1–100 nm. The brittle materials such as glasses, silicon, and germanium are machined with high-speed ultra-precision lathe to understand the material removal mechanism in different modes. When the applied depth of cut is below the critical value, it is considered to be in the ductile mode, and it can be easily machined without crack formation. Therefore, there is a significance to identify the ductile to brittle transition for these materials where the magnitude of the critical depth of cut varies depending on the properties of the components. The experiments are conducted in different atmospheres such as dry, water, methanol, octane, ethanol, and propane to form a crack free surface for the brittle materials. The machining performance of BK7 glass in the ductile mode under the methanol and ethanol conditions gives a smooth crack-free surface [1].

Generally, the single crystal silicon is frequently used in the microelectromechanical system (MEMS), where the material is finally machined to a quality product with the effect of ultra-precision grinding and polishing operation. Although the behavior of silicon material is brittle at room temperature, it is advisable to machine the silicon in the ductile mode by using a diamond turning tool. This reduces the damage caused by the brittle fracture of the ceramic material and improves the productivity of the final part [2]. The high-speed machining of nonferrous materials such as copper, aluminum, and nickel with the diamond tool is performed to evaluate the tool wear, cutting force, and surface finish. The experiments are conducted for different cutting speed such as lower speed of 150 m/min and higher speed of 4500 m/min. The rate of tool wear observed in the lower cutting speed is more than the higher cutting speed. This may be due to the reduction in the time taken for the tool engagement with the workpiece per revolution at higher speed [3]. It also reduces the chemical affinity between the tool and the workpiece interface. The diamond tool with high negative rake angle is used to finish this material in the ultra-precision accuracy. However, it provides a good surface finish to the material, but the tool wear is a major problem in machining the hard brittle material. For example, the flank and crater wear occurs obviously in the turning operation, which results in increased production cost and deteriorates the product quality.

2. Significance of diamond as a cutting tool

There is an increase in demand for preparing the plastic molds that are required for making the aspheric lens of CD optical pickup and smart lenses of the camera. The cutting edge of the tool must be sharp and free from irregular shape to process high-precision aspheric surfaces. There exists a major difference between the single crystal diamond (SCD) and the polycrystalline diamond (PCD) based on the sharpness of the tool. The cutting edges of the SCD tool are homogeneous and free from irregularities, whereas the cutting edges of the PCD tool show microscopic irregularities which lead to the removal of diamond particles. The major drawback of the SCD tool is the shorter life period compared to the PCD tool due to the abrasion wear [4, 5]. It is also used to machine the aluminum substrates to a fine mirror finish, which is used in the hard disk drive of the computer storage system. The different types of diamond cutting tools used in the ultra-precision machining are depicted in **Figure 1**.

The aspheric lenses made of glasses are used in the optical sensor for the automobile industry, where the glass lenses are pressed at high temperature in a ceramic mold. The schematic diagram of glass lens formation is depicted in **Figure 2**. These molds are produced by the ultra-precision secondary machining processes like

Figure 1.
Ultra-precision diamond cutting tools [5].

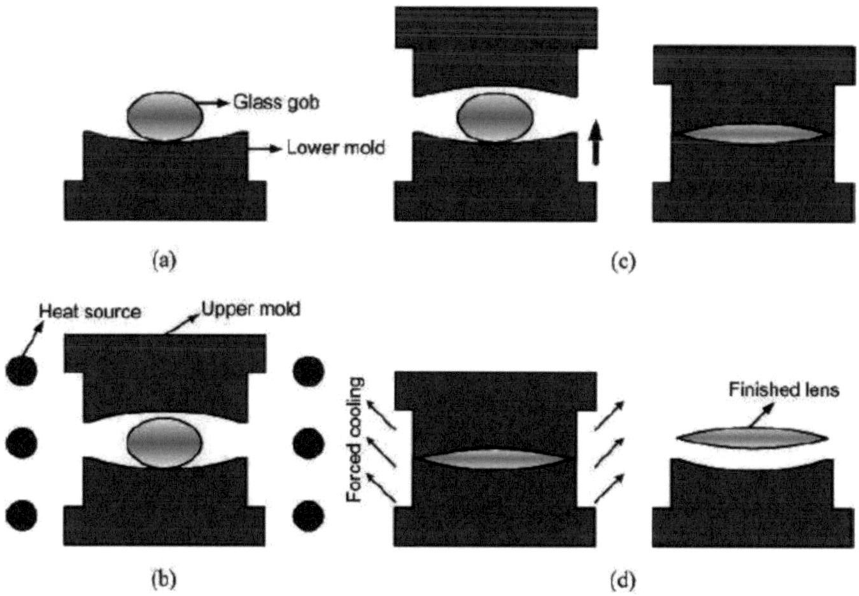

Figure 2.
Schematic diagram of glass lens formation. (a) Molds and glass gob, (b) Heating, (c) Heating and pressing, (d) Cooling and release [7].

micro-turning and micro-milling processes. While the ceramic material like tungsten carbide is initially used for making these molds, harder tools like single crystal diamond are used to shape the ceramic materials. In recent times, silicon carbide ceramic is used for making the ceramic mold instead of tungsten carbide due to the high hardness [6]. The single crystal diamond tool is not suitable to machine this type of SiC ceramic mold because of the increase in the tool wear. The damages easily occur to the SCD tool during the machining of silicon carbide ceramics due to the cleavage and the anisotropic property of the diamond tool. This scenario can be overcome by using a nano-polycrystalline diamond (NPCD) tool as an alternative to the SCD tool. Since the NPCD tool consists of fine grains of nanometer size over the length with no additional binder materials, the properties of the NPCD tool are much better than that of the SCD. The performance of the NPCD tool is improved compared to the SCD, where the tool life is longer for the NPCD. This may be due to the hardness property and thermally stable characteristics of the NCPD tool [5, 7].

Recently, plastic lenses have been produced more instead of the optical glass lenses. This may be due to the increase in mass production and reduction in the cost of the product. The different types of lenses used in the microelectronic applications are shown in **Figure 3**. Generally, the aspheric lenses are produced by the injection molding process in mass production, where the ultra-precision lathe is used as a secondary machining process to finish these lenses. This machining has the capability to control the cutting edges of the tool to 1 nm accuracy. If the cutting edge circular arc of the tool is deformed in the grinding process, it will transfer the error to the aspheric lenses and deteriorate the accuracy of the finished product. It is hard to grind the single crystal diamond to a circular arc due to the anisotropic property of the material. Another form of the lens is the hologram which is used to focus the laser light to different points, where two portions of the lens are responsible for projecting the light to the target field. In addition, the aspheric surface is considered to be a portion, and the other portion is a diffraction grating inscribed in the lens. The sharp edges and the circular profile of the

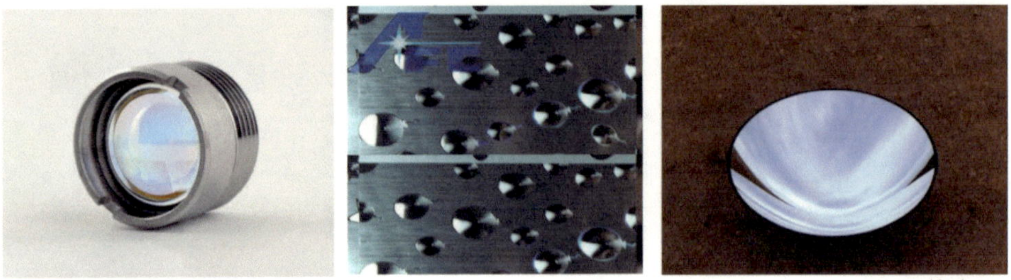

www.photonics.com www.aftech.com www.grabhad.com

(a) (b) (c)

Figure 3.
Different types of lenses (a) aspheric lens, (b) hologram lens, (c) parabolic mirror.

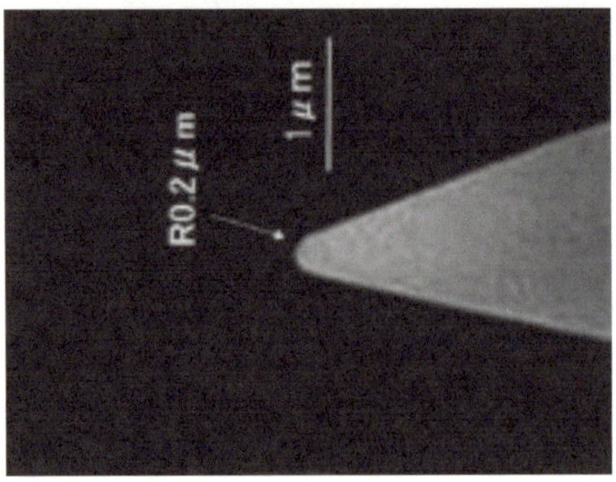

Figure 4.
Circular cutting edge of diamond tool [5].

cutting edges are the essential characteristics used to machine the hologram lens mold. Moreover, the circular cutting edge has a smaller corner radius of 0.2 μm to machine the molds in high accuracy surface as shown in **Figure 4**. The ultra-precision diamond tool is also used to slice the parabolic mirror that is available in the solid CO_2 laser to condense the light at a focal point and reflect the laser light from the mirror finished surface. The parabolic mirror produced by the copper material consists of high thermal conductivity [5].

3. Diamond-coated tools

The diamond film is coated on different substrates to improve the hardness, corrosion resistance, and wear resistance of the tool. It also removes the heat generation from the cutting edge due to the high thermal conductivity of diamond. This leads to the increase in the durability of the coated tools in the machining process. The deposition of diamond films on the substrates such as metal alloys, cemented tungsten carbide, and stainless steel has not been consistent due to the adhesion difficulty at the interface. Additionally, a chemical affinity reaction takes place between the substrate and the diamond deposited films. This enhances the formation of carbide layer when the carbon from the diamond film interacts with air particles. However, when the coating is not regular on the substrate, the diamond film peels off from the surface and reduces the tool life.

The cemented tungsten carbide with the diamond coating is used in the recent periods to enhance the tool life and the cutting performance of the dental

(a) (b) (c)

Figure 5.
Diamond coated tools (a) sintered diamond tool, (b) diamond coated (WC) tool, (c) titanium nitride interlayer tool [8].

materials as shown in **Figure 5**. The tool contains 6% cobalt metal, which is removed from the surface of the substrate by the etching process to improve the adhesion of diamond film with the tungsten carbide substrate. The alternative method to increase the adhesion between both the surfaces is by introducing a new layer of titanium nitride, which provides better adhesion and improves the diamond growth of the tool. Therefore, different types of tools such as diamond-coated (WC), diamond-embedded (sintered), and titanium nitride interlayer tools are used to investigate the drilling performance of the human tooth, glasses, and artificial tooth materials. The deposition of titanium nitride between the diamond film and the substrate performs well compared to the other tools for the orthodontic application [8].

The diamond film deposits as different layers such as monolayer, bilayer, and multilayer on the surface of the cemented tungsten carbide milling tool by the chemical vapor deposition process. The cutting performance of the milling tool is determined on the sintered zirconia ceramic, where the tribological behavior of the tool is identified by using the reciprocal tribometer. The experiments are carried out in the CNC milling machine by comparing the microcrystalline and nanocrystalline layers of the diamond tool without considering the lubrication. The monolayer deposition with nanocrystalline diamond includes a lower coefficient of friction (0.126) due to its smooth surface, the monolayer deposition with micro-crystalline diamond provides more coefficient of friction as 0.290. The monolayer diamond coated has a poor tool life due to the higher hardness of the workpiece material. The performance of the multilayer diamond-coated tool exceeds in terms of durability (3–7 times higher) compared to the monolayer and bilayer deposition [9, 10]. Hence, the small particles of diamond are embedded in the tool material for performing different micromachining operations such as turning, milling, drilling, and grinding. This enhances the tool life of the substrate by reducing the wear, built-up edges, and thermal damages to the tool surface.

4. Ductile regime machining of brittle materials

The ductile regime machining of brittle materials produces the material removal in the plastic deformation zone when the applied stress is below the critical stress of the material, which is insufficient to cause the macrocrack formation. The mechanism of material removal is identified by the occurrence of radial, lateral cracking, chipping, and pileup formation on the surface level. The morphology of machined surfaces provides the regular or irregular lay pattern, which confirmed the mode of

S. no	Materials	Fracture toughness, K_{IC} (MPa $m^{1/2}$)	Young's modulus, E (GPa)	Hardness HV (GPa)	Critical depth of cut (μm)
1	SiC	4	420	24	0.5
2	BK7 glass	0.2	81.5	5.1	0.2
3	Silica	0.75	70	6	0.1
4	Zirconia	8.05	210	13	0.97

Table 1.
Critical depth of cut of different materials [1].

deformation. The chances of both the ductile and brittle mode of deformation occur in the ceramic materials depending on the critical depth of cut. The critical depth of cut d_c (μm) is calculated by Eq. (1) given as.

$$d_c = 0.15\ (E/H)\ (K_c/H)^2 \tag{1}$$

where E is the Young modulus (GPa), H is the Vickers hardness (GPa), and K_c is the fracture toughness (MPa $m^{1/2}$) of the material [11]. When the depth is below the critical depth of cut, it leads to a ductile regime of machining, where the material removal takes place by plastic deformation. On the other hand, when the depth of cut exceeds the critical depth, it removes the material by brittle fracture. The indentation and scratch tests are conducted on different materials using a diamond tool to understand the ductile-brittle transition as listed in **Table 1**. The tool consists of a different rake angle that is used to study the quality of the surface by varying the depth of cut [12].

4.1 Scratch test behavior of brittle materials

The scratch experiment indicates the scratching of the ceramic material with a diamond grit, which is used to simulate the ceramic machining process. The diamond grit with high negative rake angle is considered to investigate the mode of material removal of brittle components, where the microscopic interaction between the tool and the workpiece is studied in detail [13]. The examiners have identified the definite method of indentation fracture mechanics to estimate the material removal mechanism of different ceramic materials without performing the machining process [14]. The material removal mechanisms of the machining process depend on the nature of the interaction of individual diamond grain with the workpiece material.

Most of the researchers carried out a single grit scratch test to identify the mechanism of material removal. The elastoplastic deformation occurs as the material removal mechanism along with brittle fracture that takes place during the scratching of hard brittle materials as shown in **Figure 6**. The scratch test has been conducted on alumina ceramic by using a single diamond grit of conical shape. The formation of plastic deformation, scalelike cracking, and cracking occurs depending on the increase in grit penetration. The chipping and cracks are produced below a certain critical depth of cut of 3 μm. The pileup area forms on both sides of the scratches at a smaller scratch depth below 3.5 μm based on the processing of microcracks. However, if the depth of cut increases more than 5 μm, the macrocrack initiates and extends within the material [15]. Therefore, the generation of cracks and the material removal rate are more, when the scratch depth increases above 5 μm. The plastic deformation not only occurs based on the undeformed chip thickness but also depends on the hydrostatic stress below the contact area of the single grit.

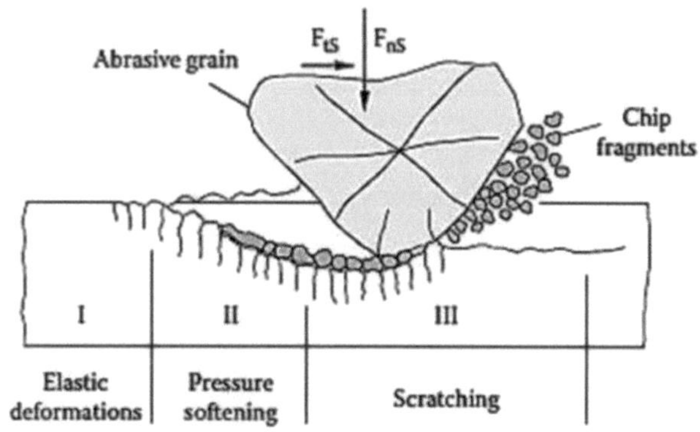

Figure 6.
Material removal process in machining of brittle materials [13].

The transition from ductile to brittle behavior is influenced by the grit shape and the material properties of the workpiece (hardness, fracture toughness, and modulus of elasticity). The mode of material removal also depends on the size of the grains which makes the structure of the material [16].

4.2 Materials having a fine grain structure

When the single grit comes in contact with the fine-grained material, it leads to the plastic deformation based on the shape of the diamond grit and the shear stress level of the material. The shear stress starts increasing depending on the penetration depth of the single grit. While the applied stress exceeds the critical stress of the material, the lateral cracks are formed parallel to the scratch direction along the sides as shown in **Figure 7**. This is associated with the easy material removal during the scratch test. The median cracks occur perpendicular to the cross section of the scratches, which extend toward the inside of the material. This causes the material strength of the workpiece to deteriorate.

4.3 Materials having a coarse grain structure

The material removal observed in the coarse structure takes place in a dissimilar way as shown in **Figure 8**, where the single grit with sharp edges plastically split the grains and the cracks are formed along the grain boundary. This leads to the break-off or chipping near the edges of the scratch. The grit with blunt edges forms a plastically deformed zone at a higher scratch depth

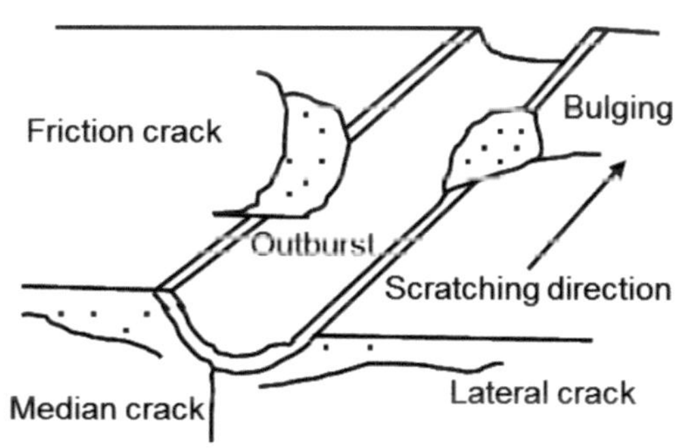

Figure 7.
Different crack system and plastic deformation in the fine-grained structure.

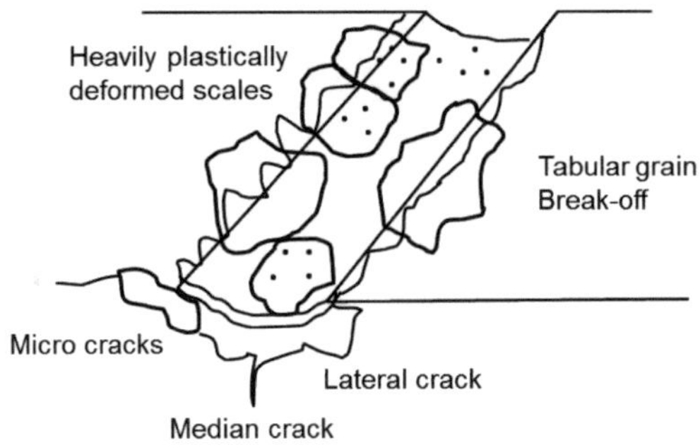

Figure 8.
Slight plastic deformation and break off in coarse grained structure.

due to the increase in shear stress. This enhances the formation of flakes around the scratch surfaces, where the material is pushed forward with more breaking chips at the edges. The single grit scratch test is usually performed to understand the interaction of a single abrasive grain with the work surface. This will enable us to understand the mode of material removal by choosing the parameters for efficient machining of the required material.

5. Tool wear of diamond in ultra-precision machining

The tool wear observed in the ultra-precision machining process is not well understood due to the incomplete studies in the past. The different types of wear mechanisms such as abrasion, adhesion, diffusion, and chemical reaction are formed on the tool surface during the machining process. The tool wear will vary depending on the machining condition and the lubrication used in the experiments. The tool wear is a significant parameter which affects the quality of the machined components and economics of the part produced. The tool wear rate is based on the properties of the work material such as strength, hardness, and fracture toughness. In order to quantify the tool wear in the turning operation, an experiment is carried out for machining the nonferrous and glass materials in the high-speed lathe. The tool wear is observed to be smooth and uniform for the nonferrous materials such as aluminum, copper, and nickel. However, the tool wear is observed to be rougher when machining the amorphous glass material. The abrasion wear occurs during the machining of glass material with the diamond tool, which is caused by the microcleavage that takes place along the crystallographic planes. The cutting edge of the diamond tool is subjected to large chipping. This is caused by the flaws and defect already present in the single crystal diamond. The increase in temperature between the tool and the workpiece interface causes thermal damage to the tool surface. The wear mechanism such as oxidation, graphitization, and diffusion occurs to the diamond tool due to the increase in the heat generation [17, 18]. The materials like glasses are difficult to machine using a sharp diamond tool due to the material property of workpiece such as low fracture toughness, amorphous structure, brittle, and fragmentation. It will also affect the material property of the diamond tool and leads to increase in the wear rapidly. A test is also carried out to evaluate the wear mechanism of the diamond tool during the machining of the glass products. The tool wear is investigated depending on the measurement of the temperature variation of the cutting tool, surface roughness of the material, and the wear zone [19]. The microchipping and cleavage are identified as the dominant wear mechanism during the machining of glass material using a diamond tool as

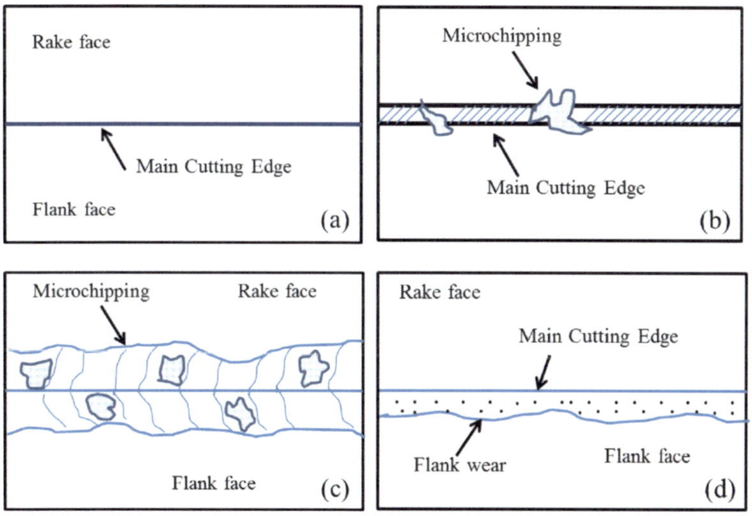

Figure 9.
Formation of wear in single point diamond turning tool (a) sharp cutting edge, (b) formation of microchipping, (c) microchipping formed over the entire length, (d) flank wear.

shown in **Figure 9a–c**. The tool life of single crystal diamond is based on the defects like nitrogen impurities present in the chemical composition. The quantity and the types of defects present in the tool material are evaluated by the infrared absorption. The result indicates that the material containing a low amount of impurities will have larger wear resistance. In contrast, the material with high B_2 aggregates will provide resistance to chipping [20].

While machining the ferrous material with the diamond tool, there is more possibility of chemical affinity between the tool and the workpiece interface. This leads to the diffusion tool wear, where the quantities of carbon content from the tool diffuse into the workpiece material. A new method of supplying nitrogen cold plasma associated with ultrasonic vibration is used to reduce the chemical affinity between the interfaces. Moreover, it also reduces the tool wear in the machining of the ferrous materials. This also supports the machining process by reducing the temperature and the surface roughness of the component [21, 22]. In another experiment, the diamond-coated tool is used to turn the graphite bar in high-speed lathe machine under the influence of both the air and nitrogen atmosphere. In addition, the oxidation forms in the presence of air, which causes wear to the cutting tool and decreases the tool life, whereas the blowing of nitrogen in the turning operation reduces the wear of the cutting tool by weakening the oxidation process. The volume of wear of the cutting edge is estimated using a coherence scanning interferometer, where the depth of wear distribution on the cutting edge is measured as a significant parameter. This chemical wear leads to rust formation on the tool edges, which is caused by the heat generated during the experiments [23].

The tool wear is observed during the machining of curved surfaces of optical components under both the ductile mode and brittle mode. In the brittle mode, the undeformed chip thickness is maintained at a certain magnitude of 900 nm, where the material removal is higher before the secondary finishing process, whereas in the ductile mode, the machining is performed at a magnitude of 90 nm undeformed chip thickness. The formation of wear on the rake face of the tool is reduced in the brittle mode. This may be due to the brittle fracture and microchipping behavior of the workpiece, where the wear land is smaller at the cutting edge. However, different types of wear such as crater wear and flank wear occur in the ductile mode as shown in **Figure 9d**, which looks smooth and uniform along the cutting edges. This shows the development of tool wear is gradual and stable throughout the machining process.

6. Monitoring of ultra-precision machining process

The online monitoring sensors are used to generate control signals during the machining process to improve the productivity of the ultra-precision manufacturing system. Some of the online monitoring sensors are given as temperature, force, power, vibration, and acoustic emission sensors, which are used to control the process effectively. The online monitoring system discusses the requirement of sensor technology in precision and ultra-precision machining process. Online sensors are used to improve the control and productivity of manufacturing systems. Therefore, the features of the sensors are correlated with the changes occurred in the output parameters of the machining conditions as shown in **Figure 10**. This involves the performance of both the tool and the workpiece material.

6.1 Machining forces

The force measurement gives detailed information about the tool wear and the surface integrity of the machined components with the help of tangential force. The power is indirectly calculated by multiplying the tangential force with the cutting velocity. The force sensor uses strain gauges in the earlier period (1950) to record the data during the machining process. Although it provides good substantial data on different operations, it is not frequently used due to the reduction in the stiffness. Recently piezoelectric quartz is used to measure the forces such as normal, tangential, and radial forces. The force sensor is placed on the machine table below the workpiece during the experiments.

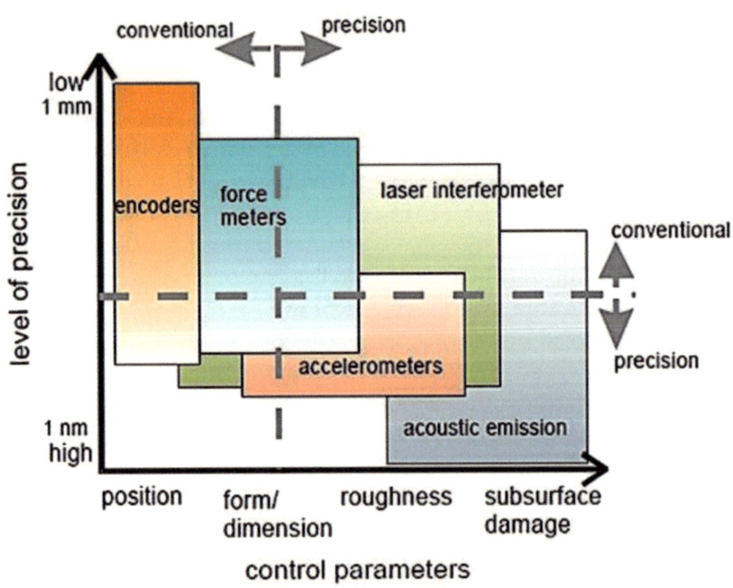

Figure 10.
Level of precision and error control parameters of sensors [19].

6.2 AE signal in ultra-precision machining

Acoustic emission (AE) is the sound waves produced when a material undergoes internal stresses as a result of mechanical loading. AE is commonly defined as the elastic waves in a material caused by the release of localized stress energy. The AE sensor is fixed to the tool or the workpiece to monitor the response of the material during the machining process. It is used to investigate the surface and subsurface damages induced by machining under different environments. The sources of the AE signals obtained from previous research works are listed as crack initiation, chipping, and material deformation. The acoustic emission signals can be generated due to the friction and wear mechanism of the scratch test, where these signals are correlated to the shape, size of abrasive grit, the coefficient of friction, and surface morphology of scratches produced. The requirements of AE sensor application in precision manufacturing and ultra-precision machining process are discussed. The online monitoring sensors are used to generate control signals to improve the control and productivity of the manufacturing systems. The sensitivity of the AE signals is related to the subsurface damages and the mode of deformation based on the type of brittle material used. The purpose of the AE sensor satisfies the corresponding requirement of machining processes such as finer finish and tighter tolerances of the workpiece material [24]. The characteristic of AE signals like root mean square (RMS) value is directly proportional to the uncut chip thickness in the diamond turning process. The sensitivity of the AE signal is related to the variation of uncut chip thickness for different diamond tools such as coated, sharp cutting edge, and worn surfaces. Therefore, an AE sensor is also used as an online monitoring system to identify the cracks of the tool in the machining process.

7. Ultra-precision grinding and polishing

The superabrasive wheels such as diamond and CBN wheels are used as ultra-precision tools to grind hard—brittle glass, ceramic, and composite materials. The problem encountered in the conventional wheel like alumina is wheel loading, which is overcome by the usage of superabrasive wheel. The diamond wheel is used to grind nonferrous material, hardened alloys, and optical glasses for the automotive application. However, the CBN wheel is used to grind hardened steels, heat-resistant alloys, and ductile metals. Although the diamond wheel is essential to grind the hardest, high-strengthened, and wear-resistant material, it resulted in self-sharpening difficulty compared to the conventional wheel. Recently diamond wheel is used to grind silicon wafer and sapphire substrate for the microelectronic application. Moreover, the conventional dressing methods like single point diamond dressing and roller dressing are supposed to remove the grain by shearing and cause damage to the grain structure of the grinding wheel. This will shorten the tool life of the grinding wheel. The dressing process of grinding operation is changed to new technologies such as electric in process dressing (ELID), laser-assisted dressing, and water jet in-process dressing to improve the dressing efficiency of the ultra-precision grinding. Ultra-precision polishing is an extension of conventional machining method, where the small particles are used to remove the material in the micron size. The material is removed in the elastic plastic deformation in that the formation of brittle fracture is avoided [25]. Mostly it is used to polish the silicon wafer for the microelectromechanical system. Initially, the wafer is chemically etched to form a flat surface, and then the chemical mechanical polishing of silicon wafer is carried out in a single-sided polishing machine as shown

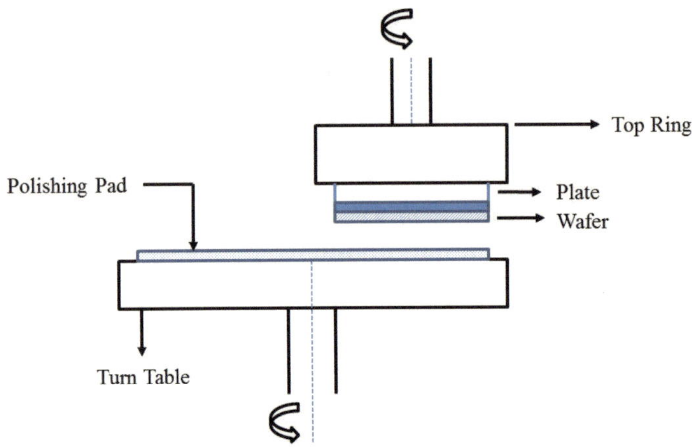

Figure 11.
Single sided ultraprecision polishing machine.

in **Figure 11**, where the wafer is fixed to a plate by the vacuum. The polishing is performed on a rotating table to which the polishing pad is attached and the load is applied to the plate through a top ring. The polishing pad is made up of synthetic leather, and the colloidal silica mixed with the alkaline solution is used as slurry to this process. This produces a mirror and smooth surface finish of the optical devices used in various applications.

8. Conclusion

There is an increase in demand for producing the high dimensional accuracy product in the optical device applications. The diamond tool is required to machine the hard and brittle materials to the specified tolerance limit. But, the occurrence of tool wear is more during the machining of molds of the optical devices, and it will deteriorate the surface finish of the part produced. This increases the cost of the production in the ultra-precision machining process. In order to overcome this scenario, the diamond film is coated to the substrate tungsten carbide to enhance the wear and corrosion resistance of the tool.

An alternative method is also used to machine this ceramic material in ductile regime, where the formation of brittle fracture is restricted. The different types of tool wear such as microchipping and crater and flank wears are observed during the machining process, and the causes responsible for these tool wears are also discussed in detail. Moreover, the monitoring of the ultra-precision machining using AE signals plays an important role in identifying the tool wear. Finally, this chapter gives an idea to understand the ductile-brittle transition of different materials and the efficient way of machining the ceramic molds using the diamond tool in ultra-precision process.

Author details

P. Suya Prem Anand
School of Mechanical Engineerig, VIT, Vellore, India

*Address all correspondence to: suyaprime@yahoo.co.in

References

[1] Ogura O, Okizaki Y. Ductile-regime turning of brittle materials by single point diamond. In: Proceeding 15th ASPE Annual Meeting; 2000. p. 58

[2] Yan J, Syoji K, Ji T. Some observations on the wear of diamond tools in ultra-precision cutting of single-crystal silicon. Wear. 2003;**255**(7-12):1380-1387. DOI: 10.1016/s0043-1648(03)00076-0

[3] Brinksmeier E, Preuss W, Riemer O, Rentsch R. Cutting forces, tool wear and surface finish in high speed diamond machining. Precision Engineering. 2017;**49**:293-304. DOI: 10.1016/j.precisioneng.2017.02.018

[4] Suzuki H, Okada M, Asai W, Sumiya H, Harano K, Yamagata Y, et al. Micro milling tool made of nano-polycrystalline diamond for precision cutting of SiC. CIRP Annals. 2017;**66**(1):93-96. DOI: 10.1016/j.cirp.2017.04.017

[5] Obata K. Single-crystal diamond cutting tool for ultra-precision processing. SEI Technical Review. 2016;**82**:83

[6] Brinksmeier E, Glaebe R, Osmer J. Surface integrity demands of high precision optical molds and realization by a new process chain. Procedia Engineering. 2011;**19**:40-43. DOI: 10.1016/j.proeng.2011.11.077

[7] Nguyen DN. FEA and Experimentally determination of applied elasticity problem for fabricating aspheric surfaces. Elasticity of Materials - Basic Principles and Design of Structures. Ezgi Günay: Intech Open; 2018. DOI: 10.5772/intechopen.79402

[8] Jackson MJ, Sein H, Ahmed W, Woodwards R. Novel diamond-coated tools for dental drilling applications. Journal of Medical Engineering and Technology. 2007;**31**(2):81-93. DOI: 10.1080/03091900500217489. PubMed PMID: 17365432

[9] Wang C, Wang X, Sun F. Tribological behavior and cutting performance of monolayer, bilayer and multilayer diamond coated milling tools in machining of zirconia ceramics. Surface and Coating Technology. 2018;**353**:49-57. DOI: 10.1016/j.surfcoat.2018.08.074

[10] Shen X, Wang X, Sun F. Fabrication and evaluation of monolayer diamond grinding tools by hot filament chemical vapor deposition method. Journal of Materials Processing Technology. 2019;**265**:1-11. DOI: 10.1016/j.jmatprotec.2018.10.001

[11] Bifano TG, Dow TA, Scattergood RO. Ductile-regime grinding: A new technology for machining brittle materials. Journal of Engineering for Industry-Transactions of the ASME. 1991;**113**:184-189. DOI: 10.1115/1.2899676

[12] Zhou M, Ngoi BKA, Zhong ZW, Chin CS. Brittle-ductile transition in diamond cutting of silicon single crystals. Materials and Manufacturing Processes. 2001;**16**(4):447-460. DOI: 10.1081/amp-100108519

[13] Marinescu ID. Handbook of Advanced Ceramics Machining. New York: CRC Press; 2006

[14] Giridhar D, Vijayaraghavan L, Krishnamurthy R. Acoustic emission response of sintered alumina zirconia composite during grooving process. NDT and E International. 2012;**46**:55-62

[15] Bi Z, Tokura H, Yoshikawa M. Study on surface cracking of alumina scratched by single point diamonds. Journal of Materials Science. 1988;**23**:3214-3224

[16] Gross TM. Deformation and cracking behavior of glasses indented with diamond tips of various sharpness. Journal of Non-Crystalline Solids. 2012;**358**(24):3445-3452. DOI: 10.1016/j.jnoncrysol.2012.01.052

[17] Zhang SJ, To S, Zhang GQ. Diamond tool wear in ultra-precision machining. The International Journal of Advanced Manufacturing Technology. 2016;**88**(1-4):613-641. DOI: 10.1007/s00170-016-8751-9

[18] Owen JD, Davies MA, Schmidt D, Urruti EH. On the ultra-precision diamond machining of chalcogenide glass. CIRP Annals. 2015;**64**(1):113-116. DOI: 10.1016/j.cirp.2015.04.065

[19] Yeo SH, Zhou M, Ngoi KA, Yap SP. Investigation of cutting temperature and tool wear in diamond cutting of glasses. Materials and Manufacturing Processes. 1999;**14**(6):875-885. DOI: 10.1080/10426919908914879

[20] Yamaguchi T, Higuchi M, Shimada S, Tanaka H, Obata K. Scientific screening of raw diamond for an ultraprecision cutting tool with high durability. CIRP Annals. 2006;**55**(1):71-74. DOI: 10.1016/s0007-8506(07)60369-2

[21] Tang Q, Yin S, Chen F, Huang S, Luo H. New technology for cutting ferrous metal with diamond tools. Diamond and Related Materials. 2018;**88**:32-42. DOI: 10.1016/j.diamond.2018.06.022

[22] Zou L, Yin J, Huang Y, Zhou M. Essential causes for tool wear of single crystal diamond in ultra-precision cutting of ferrous metals. Diamond and Related Materials. 2018;**86**:29-40. DOI: 10.1016/j.diamond.2018.04.012

[23] Hashimoto M, Kanda K, Tsubokawa T. Reduction of diamond-coated cutting tool wear during graphite cutting. Precision Engineering. 2018;**51**:186-189. DOI: 10.1016/j.precisioneng.2017.08.009

[24] Dornfeld DA, Lee Y, Chang A. Monitoring of ultraprecision machining processes. International Journal of Advanced Manufacturing Technology. 2003;**21**(8):571-578

[25] Brinksmeier E, Mutlugünes Y, Klocke F, Aurich JC, Shore P, Ohmori H. Ultra-precision grinding. CIRP Annals. 2010;**59**(2):652-671. DOI: 10.1016/j.cirp.2010.05.001

Chapter 4

Development, Properties, and Applications of CVD Diamond-Based Heat Sinks

José Vieira da Silva Neto, Mariana Amorim Fraga and Vladimir Jesus Trava-Airoldi

Abstract

Heat sink is an essential component to nanoelectronics, microelectronics, and optoelectronics applications because it allows the thermal management of devices such as integrated circuits (ICs), microelectromechanical systems (MEMSs), and graphic unit processing. There are different materials being employed for heat sink production. Among them, diamond has stood out due to its excellent chemical and physical properties. This book chapter focuses on the development, properties, and applications of CVD diamond heat sinks. It covers the basic concepts of heat conduction applied to CVD diamond as a heat sink material and its production as freestanding CVD wafers of polycrystalline CVD diamond, since the literature about this topic is extensive, giving the reader a comprehensive overview. We will comprise the use and potential widening of applications of in CVD diamond heat sink technology, providing the reader with a substantial background at the current development of solutions and new frontiers in the practical use of CVD diamond thermal management devices.

Keywords: CVD diamond, deposition, heat sinks, thermal management, freestanding diamond films

1. Introduction

Diamond is widely known as the best thermally conductive material in nature, despite having no electrical conductivity when free of dopant impurities, its unique combination of lattice configuration and strong covalent electronic bonds result in a great capacity of heat conduction through phonons. This type of heat flow also occurs in some other insulating materials, in the form of waves and wave packets [1]. Combined with its low dielectric loss, low friction coefficient, transparency for a large range of light spectrum, high hardness, and chemical inertness, diamond is certainly an exceptional material choice [2]. This set of specific properties makes diamond an ideal candidate for use in thermal management tools, especially heat management deviecs [3]. Unfortunately, there are still limiting factors for this technology, as for example, limited size substrate and impurity control besides low availability of natural diamond. In recent years, the synthesis of diamond in the laboratory has been widely studied and improved and large-scale industrial facilities were developed around two major technologies of synthetic diamond production, known as high pressure and high temperature (HPHT) and chemical vapor deposition (CVD) [4, 5]. HPHT produces single crystalline synthetic gems similar to natural diamond. However, size availability and control of impurities, such as substitutional atoms of other elements and nondiamond carbon inclusions, are limitations of this method [6]. It is not uncommon to find HPHT diamond with a yellow to brown hue, characteristic of the absorption of C-defects or nitrogen impurities [7, 8]. On the other hand, CVD diamond synthesis takes advantage of the capability of producing polycrystalline films, which present closely the same properties of single crystal diamonds, with the possibility of deposition over nondiamond substrates and large deposition areas [9]. Additionally, this method provides better control over impurities and doping once it is based on the breakdown of hydrocarbon molecules inside a controlled atmosphere chamber [10]. Currently, the CVD diamond industry is evolving fast, with larger area of deposition and in the high growth rate for single- and polycrystalline CVD diamond, but many efforts are still being made in order to achieve larger areas, thickness, and quality uniformity [11]. The main prerequisite for CVD diamond films to be applied as heat sink components is that thicker freestanding films with good uniformity should be grown through a reasonable large area [12]. Thick films are easily laser cut and polished to nanometer level smoothness of the surface. However, an additional issue is that polycrystalline films present some fragility caused by grain boundaries, an additional issue to deal with [13, 14]. In high power or high frequency electronic applications, with the increasing switching speed of components, the heat generated consequently increases, leading to higher thermal power densities to dissipate which is also affected by the size reduction of electronic components, creating a growing demand of performance for heat management components [15]. However, good thermal conductivity is not the key by itself, to an adequate heat extraction system, a good homogenous contact between heat generating component and heat sink is crucial to rapidly conduct excess heat and suppress the formation of "hot spots" which critically decreases the performance and efficiency of electronic devices [13, 16].

Since the 1960s, studies have reported the use of diamond heat sink. In 1967, Swan mounted silicon avalanche diodes on diamond to achieve a continuous power density more than twice that found on copper [17]. That same year, Dyment and D'Asaro achieved continuous operation of GaAs junction lasers at a heat sink temperature of 2000 K using diamond heat sinks [18]. In 1968, Josenhans discussed diamond as an insulating heat sink for a series combination of IMPATT diodes [19]. In a subsequent study, Decker and Schorr demonstrated that the performance of IMPATT diodes is significantly improved using diamond heat sinks [20].

Between the 1970s and 1980s, more theoretical and experimental studies were published on diamond heat sinks. In 1972, Bernick calculated the steady-state temperature profile and thermal-spreading resistance for a uniform heat source on a semi-infinite type IIa-diamond heat sink [21]. Also in 1972, Russell and Thomson described a new technique for diamond heat sink fabrication based on diamond embedded in a copper rod [22]. In 1976, Hudson measured the thermal resistivity of diamond heat-sink bonds in the form of sputtered titanium and gold films at temperatures from 1.2 to 300 K [23]. In 1977, Burgemeister showed the practical improvements obtained in the use of diamonds as heat sinks comparing the thermal resistances of samples with various kinds of metal/diamond interface measured at about 400 K with a radiation detector [24]. In 1979, Ino et al. fabricated 80 GHz band silicon d.d.r. Impatt diodes with diamond heatsinks [25].

In the 1980s, the researches on diamond for electronics continued to be motivated by the use of diamond heat sink substrates for semiconductor device applications. In 1984, Leistner fabricated a high-power silicon pnn + impatt diodes by a simple diffusion process for operation on diamond heat sinks [26]. In 1988, Doting and Molenaar developed a mathematical model of a copper-supported diamond heat sink using an iterative scheme to compute the temperature and flux distribution at the diamond-copper boundary [27]. In 1990, Csanky reported a thermal model of double-drift IMPATT diodes on diamond heat sinks. This thermal model approximates the temperature-dependent thermal conductivities of Si and diamond (Type II) by means of simple empirical formulae [28].

In early 1990s, Ramesham et al. reported a novel method of fabrication of microchannels in synthetic polycrystalline diamond thin films for heat sinking applications [29]. In 1993, Beck et al. discussed the maximum temperatures in diamond heat spreaders using the surface element method [30]. Subsequently, the enhanced performance in GaAs TUNNETT diode oscillators above 100 GHz through diamond heat sinking and power combining was reported by Eisele and Haddad [31]. An interesting study on a novel cooling system for high-power laser-diode arrays using microchannels in CVD diamond was developed by Goodson et al. They proposed a system which uses a microchannel heat sink made of chemical-vapor-deposited diamond in order to reduce the array-to-coolant thermal resistance using a simple model for the combined conduction and convection problem [32]. In 1999, Bewley et al. demonstrated a high-temperature continuous-wave laser with diamond-pressure-bond heat sinking [33].

From the 2000s to the present, an increase in the number of publications related to diamond-based heat sinks has been observed. In 2003, Labudovic and Burka developed a three-dimensional finite element model of heat transfer and residual stress within high-power laser diodes and their heat sinks. The model addresses both p-side down and p-side up laser diodes mounted on a variety of commercially available gold plated diamond heat sinks [34]. In 2006, Jessen et al. reported the operation of AlGaN/GaN high-electron mobility transistors (HEMTs) atomically attached to a CVD diamond substrate. The results demonstrated the feasibility of producing GaN-based devices on polycrystalline CVD diamond substrates to maximize heat extraction from devices operating at high power [35]. In 2008, Schneider et al. reported a study on CVD polycrystalline diamond films for heat sink applications. The measurements of dielectric strength confirmed the potential of CVD diamond for thermal management in power electronics [36]. Still in 2008, Ekimov et al. reported the thermal conductivity of diamond composites sintered under high pressures [37]. In 2010, Parashchuk et al. reported that the use of diamond heat sinks extends the drive current range of pulsed diode bars by a factor of 2–3 and enables them to operate at more than one order of magnitude [38]. In 2018, Rogalin et al. discussed several applications of diamonds in high-power optics and electronics. The main features of diamond optics of heat sinks for high-power electronics were highlighted [3].

This chapter focuses on the application of CVD diamond in heat sinks. It covers three main aspects of this technology: (i) heat transfer in diamond, (ii) CVD diamond heat sinks, and (iii) free-standing CVD diamond wafers. The first explains the mechanisms of heat transfer in diamond and correlates to its properties. The second introduces the principles and concepts related to diamond heat sinks. Finally, we briefly introduce the prospects of development and processing of CVD diamond wafers.

2. Heat transfer in diamond

Heat can be understood as the transfer of some quantity of movement between particles, whether inside a solid, liquid, or gas. This transfer can be mediated by different physical mechanisms, such as vibrational waves in the lattice of solids, known as phonons, free electrons, fluidic particles, and photons. Particularly for solids, the kinetics of heat transfer is in its majority caused by free electrons moving through the bulk material for conductors and by vibrational waves through the lattice for insulator compounds [1].

Diamond is a unique case when it comes to heat transfer, the nature of carbon covalent bonding of the structure results in a strong interaction between atoms. Carbon has four electrons in its valence band which in the fundamental energy state are distributed as $2s^2$, $2p_x^1$, $2p_y^1$, so it is natural to think that it would only form two bonds with other atoms. However, with the addition of sufficient energy, a special configuration can be achieved, promoting an electron from an s orbital to an empty p orbital, a phenomenon named sp^3 hybridization resulting in a $2s^1$, $2p_x^1$, $2p_y^1$, $2p_z^1$ distribution and allowing four bonds with other atoms. This promotes the characteristic tetrahedral organization of the atoms in the diamond lattice [39].

Upon those characteristics combined with light atomic mass, phonons inside this material lattice possess extremely high frequencies and acoustic velocities; moreover, phonon-phonon umklapp scattering[1] around room temperature is low [40], this phonon scattering being the cause of thermal resistance [41].

In dielectric crystals, thermal conductivity reaches a maximum depending on temperature, and for diamond, this maximum is reached near 100 K for pure natural type IIa single crystals, increasing at $\sim T^3$ rate from lower temperatures until this peak and then decays as the temperature continues to rise since umklapp scattering begins to activate [42].

The general approach considers an analogy from gas conductivity as follows in Eq. (1) [43]:

$$\kappa = \frac{1}{3} C V \Lambda \tag{1}$$

where κ is thermal conductivity, C is the specific heat, V is phonon velocity, and Λ is the mean free path for phonons in the solid; the mean free path is a function of many factors, and scattering of phonons can be induced by other phonons, by the walls of the crystal, grain boundaries, chemical impurities, vacancies, dislocations, and isotopes [43].

The specific heat or heat capacity per unit mass C, and thermal diffusivity D can be related to the thermal conductivity through the material mass density ρ, described in Eq. (2):

$$\kappa = \rho C D \tag{2}$$

Thermal diffusivity governs the transient response to heat input. It is often more easily measured than thermal conductivity, which governs steady-state response. Nontheless, thermal conductivity is the usually required property of interest in heat management applications [44].

With a rising temperature, thermal conductivity rises, but as there are more phonons to transport heat and as this population of phonons rises, the probability of phonon-phonon interactions also rises leading to higher probabilities of phonon decaying into other frequency phonons; this process will be dominant in any kind

[1] Umklapp Scattering: first described by Rudolf Peierls, in a general manner, umklapp scattering differs from normal phonon-phonon scattering in momentum conservation, the first does not conserve phonon momentum while normal process conserves it; Maznev and Wright discuss in-depth nuances in a more recent work, where they shed some new light on this scattering mechanisms and suggest other parameters to determine differences between these two processes; once as stated by these authors, "conservation of momentum" can cause some confusion and may be inaccurate to describe the phenomenon.

of diamond at sufficiently high T. For intermediate temperatures, scattering time τ is determined by impurities or other crystalline imperfections in diamond, such as isotope disorder. For temperatures below 100 K, only low frequency phonons are thermally excited; for this case, the dimensions of the sample are significantly more important as they determine τ [45], what explains for example, the anisotropy of κ in polycrystalline diamond films. In large, high quality diamond single crystals, phonon scattering that causes thermal resistance is predominantly caused by phonon-phonon and phonon isotope scattering [43].

In their work, Morelli et al. used an approach based on the Debye model to predict and analyze thermal conductivity of CVD diamond films produced by HFCVD and MPACVD. With this model and experimental data, they reported values ranging from 1200 to 1500 W m^{-1} K^{-1} at room temperature, reaching a peak at 220–250 K for HFCVD films, and ~1400 W m^{-1} K^{-1} for MPACVD film with a peak at 170 K [46].

Debye temperature indicates the temperature for which the occupation number of optical phonons becomes significant, indicating the crystal's highest normal mode of vibration [1, 47]. In diamond, it is 2200 K [48] and this high Debye temperature is what gives diamond low thermal resistance and low expansion coefficient.

2.1 CVD diamond thermal properties

For CVD diamond, the success as a heat sink or heat spreader device is influenced by grain size and point defect scattering of the deposited film. In addition, it is also affected by the distance from the nucleation side of the films [49–52].

The thermal properties are also influenced by the C-H bonding in the deposited layer because the process of thermal conductivity in diamond is controlled by phonon scattering rate on different kinds of defects, grain boundaries and phonon-phonon interactions. Since hydrogen decorates the defects, hydrogen concentration is a convenient indicator of imperfections in CVD diamond, as well as the isotopical nature of carbon [42]. The natural occurrence of carbon stable isotopes in diamond ^{12}C and ^{13}C is 98.9, and 1.1%, respectively; the enrichment of the isotopic ^{12}C carbon up to 99.9% has showed to enhance the thermal conductivity of diamond at room temperature by nearly 50%, and calculations made from Callaway's theoretical model shows that a 99.999% ^{12}C isotope content should enable a thermal conductivity higher than 2000 W/mK [43, 53]. **Table 1** demonstrates this variance with a comparison between common heat sink materials and natural and CVD diamond, emphasizing the variety of reported thermal conductivities for different kinds of CVD diamond films, depending on its production method and chemical nature.

Within this set of data, it is interesting to notice that some CVD diamond films with impaired quality by defects or smaller grain size present a still higher thermal conductivity at room temperature when compared to conventional metallic materials.

However, diamond is a homopolar solid. It is of primary importance to notice that thermal conductivity values change depending on which direction heat flows in CVD diamond films; this is due to a singular characteristic of this material. The CVD of diamond produces significantly high quality diamond thin films but the initial ~100 nm of grown material at the nucleation side of the film, also called coalescence layer, contains a higher density of grain boundaries due to smaller grain size what usually reduces thermal properties due to higher nondiamond carbon content; in addition, the columnar structure of films causes a sensible anisotropy of thermal conductivity, being higher in growth direction when compared to bulk crystals [12, 49, 55]. This difference between in-plane ($\kappa\|$) and perpendicular ($\kappa\perp$)

Material	Thermal conductivity (W/mK)	Thermal expansion $(10^{-6}/K)^*$	Reference
Aluminum	247	23.8	[54]
Copper	398	16.8	[54]
Diamond (natural)	2200	0.8–1.0	[49]
CVD diamond (SCD film)	790–1800 (varying with C-H bonding)	1.0–1.5	[42]
CVD diamond (MCD film)	141		[55]
	551		[56]
	950 (double layer, MPACVD)		[57]
	1000		[58]
	700–2100		[49]
	1000–1500		[59]
	1200–1500 (HFCVD) and 1400 (MPACVD)		[46]
	1000–2000 (varying with film quality)		[60]
CVD diamond (NCD film)	80–800		[51]
CVD diamond (UNCD film)	7.8–26		[56]

Values from Ref. [61].

Table 1.
Comparison between thermal conductivity and expansion coefficients of common heat sink materials, natural diamond and CVD diamond.

to growth direction is usually in the range of 10%, but can reach up to 50% [57]. It is worth emphasizing that in diamond, for room temperature, almost all heat is conducted through phonons with ~10 μm mean free path; as a consequence, polycrystalline diamond films with average grain size smaller than ~20 μm have reduced thermal conductivity for room temperature and above [52].

One of the limiting factors in the use of CVD diamond as a heat transfer layer in the industry of high power electronic devices is the direct deposition over other composition devices, for example GaAs, due to its difference in thermal expansion coefficients. The thermal mismatch often causes an accommodation-related stress of the film-substrate system and in some cases can lead to complete detachment of deposited layer, a property strongly influenced by film quality and also grain size [62, 63]. This characteristic is what probably brought the practical use to the technologies applied nowadays, the deposition over a substrate chemically and structurally compatible with diamond properties and growth conditions, etching or mechanical removal of substrate material, and proper post deposition processing for matching application requirements, detailed in the subsequent section.

3. CVD diamond heat sinks

Laboratory-grown CVD diamond has been extensively studied, not only for its exceptional set of properties but also because the CVD method exceeds limiting barriers for exploration niches in science and industry [64, 65]. This technology is based on the metastability of the crystalline phase of carbon and diamond, under certain environmental conditions. By using a mixture of hydrocarbon gas in excess of hydrogen gas under pressures below of atmospheric, with an energy source for gas dissociation and subsequent reaction, carbon condensate to solid state enabling diamond crystals growth and then multiple growth nuclei coalescence form a continuous film over a surface [66].

As a brief summary of some applications that found in CVD technique a way to apply the properties of diamond where it was either not possible or excessively expensive, we have optical windows [67, 68], cutting tools [69–71], dental application tools [72, 73], conductivity and radiation sensors [74–76], and as a special matter for this chapter, heat sinking components, such as but not limited to transistor high power devices and high power laser diodes [32, 77].

Many deposition methods have been developed since its first report in academia by Derjaguin and Fedoseev in 1975, between the most used methods to grow diamond at low pressures from hydrocarbon gas hot filament, and microwave plasma activation techniques are the ones chosen for either cost efficiency and low complexity apparatus or higher quality films and better growth rates [2, 70].

The hot filament chemical vapor deposition (HFCVD) method came along with a discovery made by Matsumoto et al. in 1982 and is based on the heating of a refractory metal filament by appling an electrical current to about 2000°C; gases dissociate by heat and filament surface reactions. Indeed, there is an interesting chemical kinetics involving hydrogen dissociation in the HFCVD; besides its activation through heat irradiated from the filaments, its dissociation also occurs at the filament surface which acts like a catalyst for the adsorption of molecular hydrogen and desorption of atomic hydrogen. When using methane as a carbon source gas, methyl radicals (CH_3) are assumed to be the major responsible specimens in the growth of CVD diamond films; the dissociation of methane into this radical is also promoted by heat and especially by atomic hydrogen chemical reactions in gas phase [66]. **Figure 1** brings a schematic design of the growth environment for the HFCVD method, where growth and filament surface regions are highlighted.

Microwave plasma-assisted chemical vapor deposition (MPACVD) was developed a year later by Kamo et al.; this technique uses a microwave radiation coupled inside a resonant cavity to activate the gas mixtures. Different geometries have been studied with the intention of reaching a closely confined plasma sphere, or in other cases, a better distributed plasma region that enables larger deposition

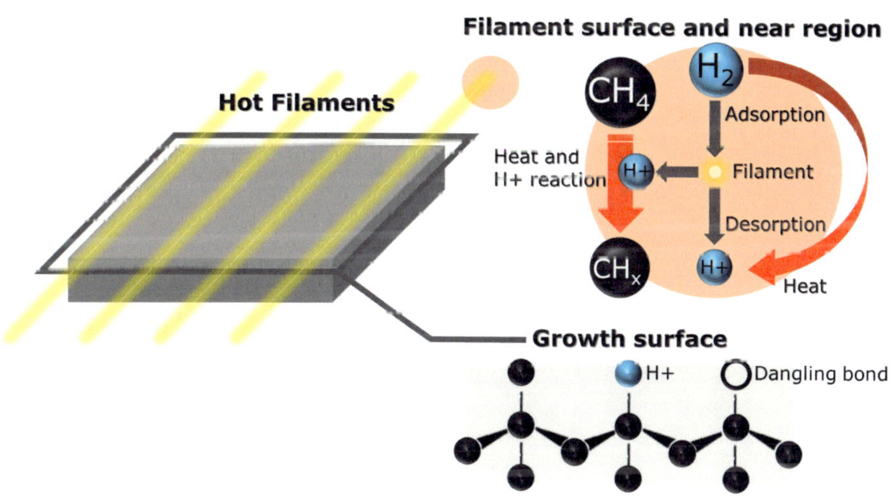

Figure 1.
HFCVD growth environment.

areas [78–80]. This technology promotes the dissociation of molecular hydrogen at considerably higher rates, once the plasma region has a larger volume and free electrons accelerated by the electric field collide with hydrogen uniformly through this volume; this also occurs to methane, the heat generated by the microwave discharge keeps the generation of free electrons and also promotes gas dissociation [81]. **Figure 2** brings a schematic design of the growth environment for the MPACVD method, where growth and plasma regions are highlighted.

The growth mechanisms involved in diamond CVD are not fully understood, but they are accepted to be similar in both HFCVD and MPACVD, starting at the point which diamond, seeded particles at nondiamond substrates or single crystalline for homoepitaxial growth, suffer reactions at its surface when exposed to gas phase with atomic hydrogen, free atomic bonds appears at the surface by atomic hydrogen abstraction, this are known as dangling bonds. Methyl radicals subsequently bind to these free sites and then atomic hydrogen keeps removing hydrogen from CH_3 molecules at the surface maintaining the sp^3 coordination, what keeps the deposition continuously occurring. Atomic hydrogen also etches graphite in a higher rate than diamond, allowing that only diamond continues to grow [65, 81].

A special issue to deal with is film surface roughness, what is generally achieved through a variety of polishing procedures detailed in a posterior section of this chapter; some deposition methods were specially developed to overcome this problem. Depending on specific growth conditions, diamond can preferentially develop different film textures varying with which crystalline plane orientation is favored or not by those conditions [82–84]. **Figure 3** shows idiomorphs or kinectic

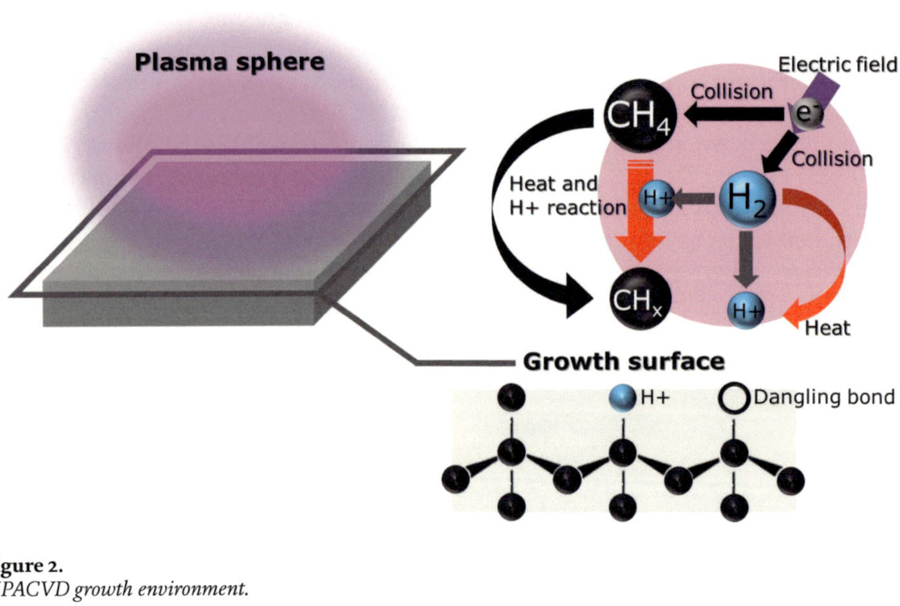

Figure 2.
MPACVD growth environment.

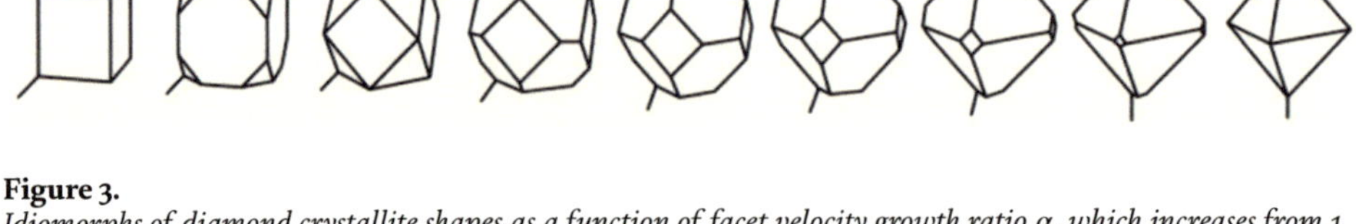

Figure 3.
Idiomorphs of diamond crystallite shapes as a function of facet velocity growth ratio α, which increases from 1 to 3 in a step of 0.25. The arrow indicates the fastest growth direction. From Ref. [85].

Wulff shapes, representations of individual crystallites more likely to grow depending on deposition conditions, that influence the growth velocities of {100} or {111} oriented faces.

This relation of growth velocity is given by the as-known α parameter that can be expressed by Eq. (3):

$$\alpha = \sqrt{3}\frac{V_{100}}{V_{111}} \quad (3)$$

where V_{100} is the velocity of {100} oriented facet growth and V_{111} is the velocity for {111} oriented facets [85]. Films with a {100} dominant morphology tend to have lower surface roughness when compared to {110} or {111} morphologies [83].

The epitaxy of diamond over iridium/YSZ (yttria-stabilized zirconia), iridium/$SrTiO_2$, or iridium/MgO substrates has successfully been adopted as a method of obtaining a crystalline diamond oriented layer using bias enhanced nucleation (BEN) to grow oriented growth nuclei, and its posterior coalescence and thickening result in high quality smooth diamond layers. In this method, diamond is nucleated by the application of a negative bias for a short period of time. This potential accelerates carbon reactive species to substrate surface, forming sp^3 and sp^2 hybridization nuclei for growth; as soon as the bias is turned off, atomic hydrogen etches nondiamond phases and only the diamond oriented particles grow until they coalesce and form a continuous oriented film [86–91].

3.1 Free-standing CVD diamond wafers

To be used as a heat transport device, diamond films have to be free of the substrate material, also known as its free standing form. There are several methods in the literature, and this section will cover the most used for fabrication of CVD diamond thermal management tools.

In the fabrication of free-standing polycrystalline CVD diamond wafers, plasma-assisted methods are preferred once they promote higher growth rates, better film quality (i.e. lower metallic inclusions), and homogeneity [92–94] however, this kind of film is also produced by the hot filament method [95].

The practical approach shown in **Figure 4** for the fabrication of CVD diamond heat sinks follows the procedures of: (1) growing diamond film over a compatible substrate, (2) removal of the substrate by mechanical procedures or chemical etching, (3) cutting and polishing of the CVD diamond top surface for reducing rough morphology in addition or not to the polishing of nucleation side to remove excessively small particles as well as the nondiamond carbon included in grain boundaries, and (4) metallization through PVD or brazing of metallic alloys or pure elements to mounting requirements, or metallization of CVD diamond surface layer by the inclusion of a dopant element, such as boron, until the metal/insulator transition is achieved.

In a recent work, Dai et al. achieved the deposition of a double layered polycrystalline diamond in order to reduce the detrimental effect of nucleation layer without removing it, on the average value of κ by a two growth step process in which diamond is deposited over silicon; then the substrate material is etched chemically by a solution of HNO3 and HF, and the freestanding film is again exposed to growth environment with nucleation side up. However, presenting a slight decrease in κ for the temperature range of 250-400 K, their solution achieved a 950 ± 14 W m^{-1} K^{-1} thermal conductivity at room temperature [57]. This study shows the deposition method versatility and the still emerging technologies we can expect on the development and use of enhanced thermal management device fabrication.

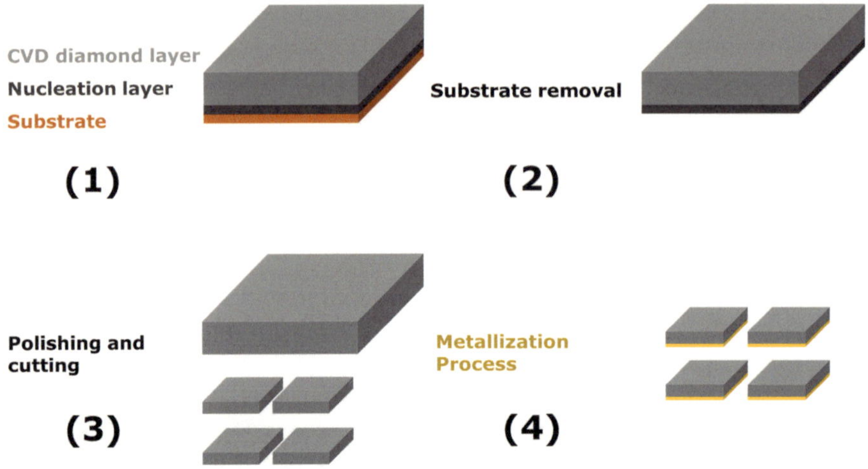

Figure 4.
Simplified scheme for production of CVD diamond heat sinks.

3.2 Post deposition processing

3.2.1 Laser cutting

The exceptional properties of CVD diamond, added to its conventional small thickness, make traditional machining and cutting methods difficult, expensive, slow, and lead to high probability of contamination at the worked surface. Lasers have been studied and applied as an inexpensive technique when compared to traditional machining, at high speed without significant surface contaminants [96, 97].

Conventionally, laser cutting of diamond is based on the transfer of energy between laser beam and workpiece; this increases the temperature at the surface of the cutting region and this heating effect can rise local temperature to above 1300°C (for a 1064 nm wavelength laser) creating an opaque zone by graphitization, resulting in a very high absorption of the incident luminous radiation. This process is practically independent of laser beam wavelength [98].

The laser cutting of diamond is affected by laser characteristics in one hand and workpiece characteristics in the other hand, for lasers wavelength, pulse energy, pulse length focal position, frequency and absorptivity at the incident radiationare the main parameters involved, the material contribution comes from chracteristics such as thermal diffusivity and presence of impurities [99]. Nd: YAG lasers at 1064 nm wavelength have less efficient coupling with diamond requiring a higher energy dose to achieve material removal by thermally dominant mechanisms. For ArF lasers with a 193 nm wavelength, more efficient optical coupling is achieved minimizing thermal load and reducing photon-affected zone to regions in close proximity to the point of beam incidence [96]. Excimer lasers have been described as the most suitable for diamond and diamond-like film processing due to higher absorption coefficient in these materials for UV phonons, as compared to visible and infrared radiation [100].

In addition to be a widely used method in diamond cutting, lasers are also an alternative in the smoothening of polycrystalline diamond films. The KrF excimer lasers (248 nm) show a high etching rate of 100 nm per pulse, with resulting roughness depending on laser pulse duration and incidence angle [101]. Watanabe et al. proposed a novel method for ultraprecision diamond polishing using UV lasers of wavelengths lower than 225 nm; their work describes a mechanism of carbon atom oxidation by UV irradiation and final mechanical removal as CO2 or CO [102].

3.2.2 Polishing

Polycrystalline diamond growth via CVD processes tends to present a granular cone-shaped rough top surface morphology that may cause poor contact between heat generating device and diamond piece; this is also combined with the effect of reduced grain size and consequent increase of grain boundaries and nondiamond phases present in the nucleation region of the film severely decreasing thermal conductivity. Dielectric surface roughness can also aggravate the conductor roughness due to conformal nature of the metallization process. The thinning and polishing of the diamond piece is then of great importance to reduce the thermal conductivity of the affected region and smoothing of the film surface to promote better contact between pieces [12, 59].

Traditional abrasive polishing methods for natural diamond are not suitable for CVD diamond films because diamond exhibits nearly a factor of 600 of polishing rate difference depending on the direction it is being polished. In a polycrystalline film, the mix of present planes limits the material removal rate to the slowest direction. A method used for the efficient polishing of CVD diamond films is the thermochemical polishing, which consists in the friction of diamond workpiece with a heated iron disk; this takes advantage of the high solubility of carbon in pure iron, 170 mg cm^{-3} at eutetic temperature and also by high carbon diffusion rate [103, 104].

Another promising method called dynamic friction polishing can be applied to both single and polycrystalline diamond polishing due to its high efficiency and flexible environment (free of abrasives). Furthermore, it is performed at room temperature and open atmosphere. The use of this polishing method on CVD diamond thin films is sensitive to two factors: (i) the applied pressure during polishing, which in excess causes films to crack and (ii) lower pressures reduce material removal rate and time of polishing and long times induce higher thermal stress [14].

An alternative method for polishing is the planarization-by-filling, which is a method that offers a simple and conventional approach to the reduction of surface roughness without any kind of surface polishing. This technique consists basically in filling of the growth surface microcavities by adding a polymer, glass, diamond-filled glass, or similar material. The thin overlayer applied can also be polished to adequate thickness and expose some of the diamond surface [105].

3.2.3 Metallization

The process of adding a metallic layer to CVD diamond films, metallization, are conducted by several different systems, most employing a refractory metal adhesion layer which when deposited under appropriate conditions, forms an interfacial refractory metal-carbide phase. Some metallization systems are based on processing noble metal surface chemistry or ion beam surface chemistry to provide a "near surface" chemically and structurally modified diamond suitable for non-refractory bonding metallurgies [59].

The need of a metal layer is dependent on application, diamond heat management devices are often metal coated for grounding, signal and power connections, and if that is the case, plating is preferred over thermal evaporation or sputtering because its simple and less costly [105].

The adhesion of the metallic layer is of paramount importance. The conductive metals like gold or copper are not suitable for direct deposition over CVD diamond due to diamond chemical inertness; then these metals platings require the application of other metal seed layers. Generally, these seed layers are of refractory metals due to their strong adhesion to diamond forming carbide complexes when exposed to annealing process at high temperatures (700–900°C). It is important to emphasize that the previously mentioned roughness of diamond can cause serious blistering of the plated metal if not made under proper conditions [105, 106].

4. Conclusion

Diamond and its laboratrial synthesis have been extensively studied and improved in the last decades. As diamond becomes more feasible as an engineering material, many industrial and technological uses begin to be economically attractive. This chapter aimed to bring an overview of current developments and remaining challenges on CVD diamond heat sink production and applications; CVD diamond polycrystalline films already reached all the properties of perfect IIa single crystal diamonds. Despite their first run in the market by the early 1990s, some technical issues still remain such as the rough surface obtained at regular growth rates, defects, and grain boundaries and intrinsic drawbacks on the CVD of diamond. Taking this into account, CVD diamond is still a better heat management material when compared to conventionally applied metals, and can be used in different levels of quality and physical properties to match specific application demands.

Acknowledgements

This research was supported by the São Paulo Research Foundation FAPESP (processes 12/15857-1 and 14/18139-8) and CAPES. M. A. Fraga acknowledges the support from CNPq (Process 421317/2018-3).

Conflict of interest

The authors declare that there are no conflicts of interest.

Author details

José Vieira da Silva Neto[1]*, Mariana Amorim Fraga[2] and Vladimir Jesus Trava-Airoldi[1]

1 National Institute for Space Research (INPE), São José dos Campos, Brazil

2 Universidade Federal de São Paulo, São José dos Campos, São Paulo, Brazil

*Address all correspondence to: jvneto.ifsp@gmail.com

References

[1] Kittel C. Introduction to Solid State Physics. 2nd ed. John Wiley & Sons; 1956

[2] May PW. Diamond thin films: A 21st century material. Philosophical Transactions of the Royal Society A. 2000:358

[3] Rogalin V, Krymskii M, Krymskii K. Several applications of diamonds in high-power optics and electronics. Journal of Communications Technology and Electronics. 2018;**63**(11):1326-1334

[4] Bundy EP, Hall HT, Wentorf RJ Jr. Man-made diamonds. Nature. 1955;**176**(4471):51-55

[5] Angus JC. Diamond synthesis by chemical vapor deposition: The early years. Diamond and Related Materials. 2014;**49**:77-86

[6] Hei LF et al. Interface features of the HPHT Ib substrate and homoepitaxial CVD diamond layer. Diamond and Related Materials. 2016;**69**:33-39

[7] Kazuchits NM et al. Comparison of HPHT and LPHT annealing of Ib synthetic diamond. Diamond and Related Materials. 2019;**91**:156-164

[8] Willems B, Tallaire A, Achard J. Optical study of defects in thick undoped CVD synthetic diamond layers. Diamond and Related Materials. 2014;**41**:25-33

[9] Kumar Mallik A, Bhar R, Bysakh S. An effort in planarising microwave plasma CVD grown polycrystalline diamond (PCD) coated 4in. Si wafers. Materials Science in Semiconductor Processing. 2016;**43**:1-7

[10] Achard J et al. High quality MPACVD diamond single crystal growth: High microwave power density regime. Journal of Physics D: Applied Physics. 2007;**40**(20):6175

[11] Weng J et al. Deposition of large area uniform diamond films by microwave plasma CVD. Vacuum. 2018;**147**:134-142

[12] Jin S, Mavoori H. Processing and properties of CVD diamond for thermal management. Journal of Electronic Materials. 1998;**27**(11):1148-1153

[13] Malshe A et al. A review of techniques for polishing and planarizing chemically vapor-deposited (CVD) diamond films and substrates. Diamond and Related Materials. 1999;**8**(7):1198-1213

[14] Feng HB, Chen YQ, Zhang LC. Polishing of CVD Diamond Wafers and Films in Key Engineering Materials. Trans Tech Publications; 2013

[15] Chen Y-J, Young T-F. Thermal stress and heat transfer characteristics of a Cu/diamond/Cu heat spreading device. Diamond and Related Materials. 2009;**18**(2-3):283-286

[16] Han Y et al. Enhancement of hotspot cooling with diamond heat spreader on Cu microchannel heat sink for GaN-on-Si device. IEEE Transactions on Components, Packaging and Manufacturing Technology. 2014;**4**(6):983-990

[17] Swan CB. Improved performance of silicon avalanche oscillators mounted on diamond heat sinks. Proceedings of the IEEE. 1967;55(9):1617-1618

[18] Dyment JC, D'Asaro LA. Continuous operation of GaAs junction lasers on diamond heat sinks at 200°K. Applied Physics Letters. 1967;**11**(9):292-294

[19] Josenhans JG. Diamond as an insulating heat sink for a series combination of IMPATT diodes. Proceedings of the IEEE. 1968;**56**:762-763

[20] Decker DR, Schorr AJ. High-power IMPATT diodes on diamond heat sinks. IEEE Transactions on Electron Devices. 1970;17(9):739-743

[21] Bernick RL. Temperature profile and thermal-spreading resistance for a semi-infinite type-IIa-diamond heat sink. Electronics Letters. 1972;**8**(7):180-181

[22] Russell EM, Thomson I. Embedded diamond heat sinks for avalanche diodes. Proceedings of the IEEE. 1972;**60**(8):1014-1015

[23] Hudson PRW. The thermal resistivity of diamond heat-sink bond materials. Journal of Physics D: Applied Physics. 1976;**9**(2):225-232

[24] Burgemeister EA. Thermal resistance at metal/diamond interfaces in relation to the mounting of microwave diodes. Journal of Physics D: Applied Physics. 1977;**10**(14):1923-1930

[25] Ino M et al. 80 GHz silicon diamond-heatsink impatt diodes. Electronics Letters. 1979;**15**(1):2-3

[26] Leistner D. Single drift impatt diodes on diamond heat sinks for W-band frequencies. IEE Proceedings I: Solid-State and Electron Devices. 1984;**131**(2):56-58

[27] Doting J, Molenaar J. Isotherms in diamond heat sinks, non-linear heat transfer in an excellent heat condition. In: Fourth Annual IEEE Semiconductor Thermal and Temperature Measurement Symposium. 1988

[28] Csanky G. Reliability critical thermal model for double-drift IMPATT diodes on diamond heat sinks. Quality and Reliability Engineering International. 1990;**6**(2):73-84

[29] Ramesham R et al. Fabrication of microchannels in synthetic polycrystalline diamond thin films for heat sinking applications. Journal of the Electrochemical Society. 1991;**138**(6):1706-1709

[30] Beck JV, Osman AM, Lu G. Maximum temperatures in diamond heat spreaders using the surface element method. Journal of Heat Transfer. 1993;**115**(1):51-57

[31] Eisele H, Haddad GI. Enhanced performance in GaAs TUNNETT diode oscillators above 100 GHz through diamond heat sinking and power combining. IEEE Transactions on Microwave Theory and Techniques. 1994;**42**(12):2498-2503

[32] Goodson KE, Kurabayashi K, Pease RFW. Improved heat sinking for laser-diode arrays using microchannels in CVD diamond. IEEE Transactions on Components, Packaging, and Manufacturing Technology: Part B. 1997;**20**(1):104-109

[33] Bewley WW et al. High-temperature continuous-wave 3-6.1 μm "W" lasers with diamond-pressure-bond heat sinking. Applied Physics Letters. 1999;**74**(8):1075-1077

[34] Labudovic M, Burka M. Heat transfer and residual stress modeling of a diamond film heat sink for high power laser diodes. IEEE Transactions on Components and Packaging Technologies. 2003;**26**(3):575-581

[35] Jessen G et al. AlGaN/GaN HEMT on diamond technology demonstration. in Compound Semiconductor Integrated Circuit Symposium Proceedings. 2006

[36] Schneider H et al. Study of CVD diamond films for thermal management in power electronics. In: 2007 European Conference on IEEE Power Electronics and Applications. 2007

[37] Ekimov E et al. Thermal conductivity of diamond composites sintered under high pressures.

Diamond and Related Materials. 2008;**17**(4-5):838-843

[38] Parashchuk VV et al. Improving the efficiency of high-power diode lasers using diamond heat sinks. Quantum Electronics. 2010;**40**(4):301

[39] Atkins P, Atkins PW, de Paula J. Atkins' Physical Chemistry. Oxford: OUP; 2014

[40] Ward A et al. Ab initio theory of the lattice thermal conductivity in diamond. Physical Review B. 2009;**80**(12):125203

[41] Maznev A, Wright O. Demystifying umklapp vs normal scattering in lattice thermal conductivity. American Journal of Physics. 2014;**82**(11):1062-1066

[42] Sukhadolau A et al. Thermal conductivity of CVD diamond at elevated temperatures. Diamond and Related Materials. 2005;**14**(3-7):589-593

[43] Anthony T et al. Thermal diffusivity of isotopically enriched C12 diamond. Physical Review B. 1990;**42**(2):1104

[44] Graebner J. Measurements of specific heat and mass density in CVD diamond. Diamond and Related Materials. 1996;**5**(11):1366-1370

[45] Callaway J. Model for lattice thermal conductivity at low temperatures. Physical Review. 1959;**113**(4):1046

[46] Morelli DT, Hartnett TM, Robinson CJ. Phonon-defect scattering in high thermal conductivity diamond films. Applied Physics Letters. 1991;**59**(17):2112-2114

[47] Debye P. Zur theorie der spezifischen wärmen. Annalen der Physik. 1912;**344**(14):789-839

[48] Varshni YP. Temperature dependence of the energy gap in semiconductors. Physica. 1967;**34**(1):149-154

[49] Graebner J et al. Anisotropic thermal conductivity in chemical vapor deposition diamond. Journal of Applied Physics. 1992;**71**(11):5353-5356

[50] Wörner E et al. Thermal conductivity of CVD diamond films: High-precision, temperature-resolved measurements. Diamond and Related Materials. 1996;**5**(6-8):688-692

[51] Anaya J et al. Thermal conductivity of ultrathin nano-crystalline diamond films determined by Raman thermography assisted by silicon nanowires. Applied Physics Letters. 2015;**106**(22):223101

[52] Simon RB et al. Effect of grain size of polycrystalline diamond on its heat spreading properties. Applied Physics Express. 2016;**9**(6):061302

[53] Wei L et al. Thermal conductivity of isotopically modified single crystal diamond. Physical Review Letters. 1993;**70**(24):3764

[54] Callister WD, Rethwisch DG. Materials Science and Engineering: An Introduction. 8th ed. Wiley; 2009

[55] Nazari M et al. Optical characterization and thermal properties of CVD diamond films for integration with power electronics. Solid-State Electronics. 2017;**136**:12-17

[56] Shamsa M et al. Thermal conductivity of nitrogenated ultrananocrystalline diamond films on silicon. Journal of Applied Physics. 2008;**103**(8):083538

[57] Dai B et al. Thermal conductivity of free-standing CVD diamond films by growing on both nuclear and growth sides. Diamond and Related Materials. 2017;**76**:9-13

[58] Ono A et al. Thermal conductivity of diamond films synthesized by

microwave plasma CVD. Japanese Journal of Applied Physics. 1986;**25**(10A):L808

[59] Fabis PM. The processing technology and electronic packaging of CVD diamond: A case study for GaAs/CVD diamond plastic packages. Microelectronics Reliability. 2002;**42**(2):233-252

[60] Coe S, Sussmann R. Optical, thermal and mechanical properties of CVD diamond. Diamond and Related Materials. 2000;**9**(9-10):1726-1729

[61] Railkar TA et al. A critical review of chemical vapor-deposited (CVD) diamond for electronic applications. Critical Reviews in Solid State and Materials Sciences. 2000;**25**(3):163-277

[62] Berry B et al. Internal stress and elasticity of synthetic diamond films. Applied Physics Letters. 1990;**57**(3):302-303

[63] Abreu C et al. Grain size effect on self-mated CVD diamond dry tribosystems. Wear. 2005;**259**(1-6):771-778

[64] May PW. Diamond thin films: A 21st-century material. Philosophical Transactions of the Royal Society of London. Series A: Mathematical, Physical and Engineering Sciences. 2000;**358**(1766):473

[65] May PW, Mankelevich YA. From ultrananocrystalline diamond to single crystal diamond growth in hot filament and microwave plasma-enhanced CVD reactors: A unified model for growth rates and grain sizes. Journal of Physical Chemistry C. 2008;**2008, 112**:12432-12441

[66] Gracio J, Fan Q, Madaleno J. Diamond growth by chemical vapour deposition. Journal of Physics D: Applied Physics. 2010;**43**(37):374017

[67] Campos R et al. Development of nanocrystalline diamond windows for application in synchrotron beamlines. Vacuum. 2013;**89**:21-25

[68] Dodson J et al. Single crystal and polycrystalline CVD diamond for demanding optical applications. In: Window and Dome Technologies and Materials XII. International Society for Optics and Photonics; 2011

[69] Cabral G et al. A study of diamond film deposition on WC–Co inserts for graphite machining: Effectiveness of SiC interlayers prepared by HFCVD. Diamond and Related Materials. 2008;**17**(6):1008-1014

[70] Schäfer L, Höfer M, Kröger R. The versatility of hot-filament activated chemical vapor deposition. Thin Solid Films. 2006;**515**(3):1017-1024

[71] Arumugam PU, Malshe AP, Batzer SA. Dry machining of aluminum–silicon alloy using polished CVD diamond-coated cutting tools inserts. Surface and Coatings Technology. 2006;**200**(11):3399-3403

[72] Trava-Airoldi VJC, Evaldo J, Bosco ED, Leite NF, Valera MC, Pena AFV, Baranauskas V. CVD Diamond Burrs for Odontological and Related Uses. 1998

[73] Sein H et al. Comparative investigation of smooth polycrystalline diamond films on dental burs by chemical vapor deposition. Journal of Materials Engineering and Performance. 2006;**15**(2):195-200

[74] Adam W et al. Performance of irradiated CVD diamond micro-strip sensors. Nuclear Instruments and Methods in Physics Research Section A: Accelerators, Spectrometers, Detectors and Associated Equipment. 2002;**476**(3):706-712

[75] Schmid G et al. A neutron sensor based on single crystal CVD

diamond. Nuclear Instruments and Methods in Physics Research Section A: Accelerators, Spectrometers, Detectors and Associated Equipment. 2004;**527**(3):554-561

[76] Mazzeo G et al. Deep UV pulsed laser monitoring by CVD diamond sensors. Sensors and Actuators A: Physical. 2004;**113**(3):277-281

[77] Seelmann-Eggebert M et al. Heat-spreading diamond films for GaN-based high-power transistor devices. Diamond and Related Materials. 2001;**10**(3-7):744-749

[78] Gu Y et al. Microwave plasma reactor design for high pressure and high power density diamond synthesis. Diamond and Related Materials. 2012;**24**:210-214

[79] Hemawan K et al. Improved microwave plasma cavity reactor for diamond synthesis at high-pressure and high power density. Diamond and Related Materials. 2010;**19**(12):1446-1452

[80] Vikharev AL et al. Multimode cavity type MPACVD reactor for large area diamond film deposition. Diamond and Related Materials. 2018;**83**:8-14

[81] Kasu M. Diamond epitaxy: Basics and applications. Progress in Crystal Growth and Characterization of Materials. 2016;**62**(2):317-328

[82] Choi IH et al. Growth of highly oriented diamond films by the MPCVD technique using $CO–H_2$, $CH_4–H_2$ and $CH_4–N_2–H_2$ gas mixtures. Diamond and Related Materials. 2004;**13**(4-8):574-580

[83] Silva F et al. Control of an MPACVD reactor for polycrystalline textured diamond films synthesis: Role of microwave power density. Diamond and Related Materials. 1996;**5**(3):338-344

[84] Gicquel A et al. CVD diamond films: From growth to applications. Current Applied Physics. 2001;**1**(6):479-496

[85] Smereka P et al. Simulation of faceted film growth in three dimensions: Microstructure, morphology and texture. Acta Materialia. 2005;**53**(4):1191-1204

[86] Lee KH et al. Epitaxy of iridium on $SrTiO_3$/Si(001): A promising scalable substrate for diamond heteroepitaxy. Diamond and Related Materials. 2016;**66**:67-76

[87] Ohtsuka K et al. Epitaxial growth of diamond on iridium. Japanese Journal of Applied Physics. 1996;**35**(8B):L1072

[88] Saito T et al. Epitaxial nucleation of diamond on an iridium substrate by bias treatment, for microwave plasma-assisted chemical vapor deposition. Diamond and Related Materials. 1998;**7**(9):1381-1384

[89] Schreck M et al. Domain formation in diamond nucleation on iridium. Diamond and Related Materials. 2003;**12**(3-7):262-267

[90] Gsell S et al. A route to diamond wafers by epitaxial deposition on silicon via iridium/yttria-stabilized zirconia buffer layers. Applied Physics Letters. 2004;**84**(22):4541-4543

[91] Bauer T et al. Growth of epitaxial diamond on silicon via iridium/$SrTiO_3$ buffer layers. Diamond and Related Materials. 2005;**14**(3-7):314-317

[92] Lee J-K et al. Free-standing diamond wafers deposited by multi-cathode, direct-current, plasma-assisted chemical vapor deposition. Diamond and Related Materials. 2000;**9**(3-6):364-367

[93] Aleksov A et al. Diamond-based electronics for RF applications. Diamond and Related Materials. 2004;**13**(2):233-240

[94] Zuo S et al. Investigation of diamond deposition uniformity and quality for freestanding film and substrate applications. Diamond and Related Materials. 2008;**17**(3):300-305

[95] Ohmagari S et al. Characterization of free-standing single-crystal diamond prepared by hot-filament chemical vapor deposition. Diamond and Related Materials. 2014;**48**:19-23

[96] Fabis PM. Laser machining of CVD diamond: Chemical and structural alteration effects. Surface and Coatings Technology. 1996;**82**(3):320-325

[97] Riva R et al. Cutting and drilling of CVD diamond using a copper vapor laser. In: High-Power Lasers: Applications and Emerging Applications. International Society for Optics and Photonics; 1996

[98] Migulin VV, Ralchenko VG, Baik Y-J. Oxygen-assisted laser cutting and drilling of CVD diamond. In: Lasers in Synthesis, Characterization, and Processing of Diamond. Vol. 3484. SPIE; 1998. p. 5

[99] Boutinguiza M et al. Cutting of CVD diamond by optical fibre guided Nd:YAG laser. International Congress on Applications of Lasers & Electro-Optics. 2003;**2003**(1):525

[100] Kononenko TV et al. Excimer laser etching of diamond-like carbon films: Spalling effect. Applied Surface Science. 1995;**86**(1):234-238

[101] Pimenov SM et al. UV laser processing of diamond films: Effects of irradiation conditions on the properties of laser-treated diamond film surfaces. Diamond and Related Materials. 1993;**2**(2):291-297

[102] Watanabe J, Touge M, Sakamoto T. Ultraviolet-irradiated precision polishing of diamond and its related materials. Diamond and Related Materials. 2013;**39**:14-19

[103] Hickey CF et al. Polishing of filament-assisted CVD diamond films. San Diego, '91. Vol. 1534. SPIE 10; 1991

[104] Zaitsev AM et al. Thermochemical polishing of CVD diamond films. Diamond and Related Materials. 1998;**7**(8):1108-1117

[105] Brown WD et al. State-of-the-art synthesis and post-deposition processing of large area CVD diamond substrates for thermal management. Surface and Coatings Technology. 1996;**86-87**:698-707

[106] Katz A et al. Au-Sn/W and Au-Sn/Cr metallized chemical vapor deposited diamond heat sinks for InP laser device applications. Journal of Applied Physics. 1994;**75**(1):563-567

Chapter 5

Intrinsic Carrier Parameters and Optical Carrier Injection Method in High-Purity Diamonds

Ikuko Akimoto and Nobuko Naka

Abstract

Diamond attracts increasing attentions as a semiconductor, since high-purity synthesized diamonds have become commercially available in these decades. For appropriate design of any devices, the basic carrier transport parameters should be known. However, it has been difficult to determine carrier parameters in diamond, because the controlled doping and Ohmic contact formation have been hard to achieve. In this chapter, a modern experimental method to measure basic carrier parameters, such as the effective mass, scattering times, and mobility of intrinsic diamonds, is introduced. The method, i.e., nanosecond time-resolved cyclotron resonance (TRCR), is applicable to optically injected carriers in intrinsic diamonds without wire connection. Following the key technique of optical carrier injection, detailed analysis methods for the cyclotron resonance spectra are introduced. The extracted basic parameters of diamond are summarized in comparison to those of silicon and germanium in the same group-IV semiconductor family. This is worthy for triggering further ideas in application-oriented researches using widespread materials.

Keywords: effective mass, scattering time, mobility, cyclotron resonance, optical carrier injection

1. Introduction

Diamond has been an attractive semiconductor in the fields of power electronics [1], valleytronics [2], optoelectronics [3, 4], and quantum information technology [5, 6] in recent years. Such application-oriented studies have been arising from the outstanding values of breakdown voltage, thermal conductivity, carrier mobil ity, and spin relaxation time in a diamond. This direction is accelerated due to the progress of crystal growth technique by the chemical vapor deposition (CVD) in these decades [7], by which a highly pure diamond becomes commercially available.

To design diamond-based devices, the knowledge of transport parameters, such as effective mass, scattering time, and drift mobility, is necessary. The effective mass is an important parameter in the band theory of a semiconductor, governing the transport properties, density of states, and the phase boundary of high-density carriers. The drift mobility involving the values of the effective mass and scattering time is a direct index of carrier transport. However, in the past when only natural crystals or impurity-rich synthesized crystals were available, limited information about intrinsic carrier properties had been reported [8–14]. This historical situation is in contrast to the current materials, silicon, and germanium. In silicon or germanium, a cyclotron resonance method played a significant importance to determine the effective masses for doped crystals under activation by light at low temperature in the 1950s [15, 16]. Such accurate measurements at low temperature had been impossible in diamond due to deep dopant states in the wide energy bandgap. Therefore, most of previous measurements were performed at temperatures higher than 80 K, where a carrier transport was limited by phonon scatterings. A part of anisotropic hole masses were obtained from unresolved spectra at higher temperature than 300 K [11, 12]. Information on the electron was much less, because most of the semiconducting diamond was of *p*-type.

Recently, measurements of time-of-flight (TOF) transport [2, 10, 17, 18] and optical transient grating [19, 20] have been performed with the highly pure crystals. However, the measured carrier mobility varied from sample to sample depending on the surface termination condition, the crystal supplier, and experimental conditions. A high-density injection under high electric field, the space charge-limited transport condition under higher dopant concentration, and non-Ohmic contact caused extrinsic effects on transport behaviors. To clarify intrinsic carrier properties in a diamond, a measurement should be achieved at low temperature under a low carrier density.

In this chapter, our recent experimental contributions to clarify the basic and intrinsic carrier parameters in a diamond will be introduced [21–27]. The measurement has been performed by a time-resolved cyclotron resonance method under optical carrier injection in pure diamond crystals. The concept of our measurement is shown in **Figure 1**: carriers are injected optically with ultra-violet laser pulses through the band-to-band transition or exciton creation with an assistance of phonon emission/absorption in the indirect band structure (**Figure 1a**). Although the created exciton is an electrically neutral binding state of electron and hole, free charge carriers are dissociated from excitons via two-body collision of excitons or thermal dissociation as described in Section 3.2. During the long lifetime of the free carriers in the indirect band structure, we can observe the cyclotron resonance under the external magnetic field (**Figure 1b**). Keys to realize our measurements in an intrinsic semiconductor diamond are the optical carrier injection technique and using of highly pure diamond. As in the case of pure silicon [28], which had been applied to a light-triggered thyristor as a successful power device, optical carrier injection is a promising technique to control a carrier density by a sophisticated way. The spectroscopic way of the optical carrier injection in a diamond at device-operating temperature as well as at low temperature will also be introduced.

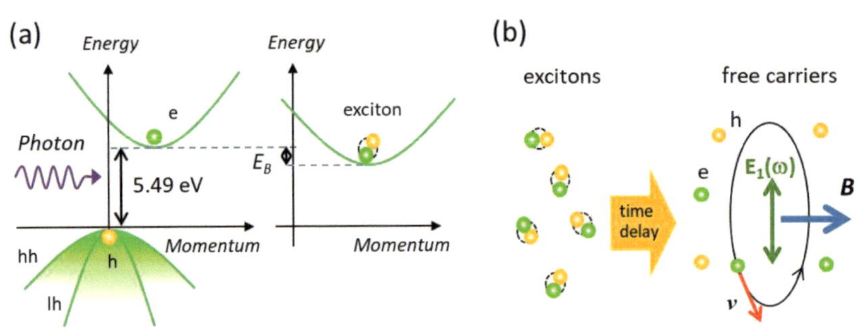

Figure 1.
Schematic experimental concept. (a) Energy diagrams of valence and conduction bands for indirect semiconductor diamond and exciton band and (b) cyclotron motion of optically injected free carriers converted from excitons.

2. Experimental method

The time-resolved cyclotron resonance (TRCR) method was performed for optically injected transient carriers in high-purity diamond crystals in X band (microwave frequency at ν = 9.6 GHz) with a time resolution of a few nanoseconds. The well-resolved CR spectrum can be observed under the condition, $\omega\tau \gg 1$, where $\omega = 2\pi\nu$ is the angular frequency of applied microwave and τ is the carrier scattering time. As the ω is selected by a used equipment, it is necessary to make τ longer by lowering the temperature at cryogenic ones and using pure crystals in order to determine the effective mass in good accuracy.

Highly pure diamond crystals of type-IIIa grown by the CVD method were used. A typical concentration of nitrogen and boron atoms was less than 5 and 1 ppb, respectively ([N] < 9×10^{14} cm^{-3}, [B] < 2×10^{14} cm^{-3}). For crystals of higher impurity concentration, it was difficult to obtain the TRCR spectrum at 10 K because of the broader spectral width due to the higher carrier scattering rate. A typical crystal dimensions were of $3 \times 2 \times 0.5$ mm^3 with the largest plane of the crystalline (001). A crystal was attached on a 2×8 mm^2 face of a right-angle prism by a small amount of vacuum grease for better coupling with the optical excitation (see **Figure 2a**). The sample was mounted in a dielectric microwave cavity (Bruker, MD5W1, TE_{011}) that is developed for the pulsed electron paramagnetic resonance (EPR) in X band with a high filling factor, in which a microwave's electric field packed in a round mode resonates with the cyclotron motion of free carriers under an external magnetic field.

The sample was irradiated by 5-ns pulses at wavelength selected in the range from 219.4 to 226.4 nm at low temperatures or from 219.1 to 235.6 nm at room temperature from an optical parametric oscillator (Spectra Physics, MOPO with frequency doubler option) pumped by THG of a Nd:YAG laser. Temporal responses of continuous microwave power were measured in a quadrature

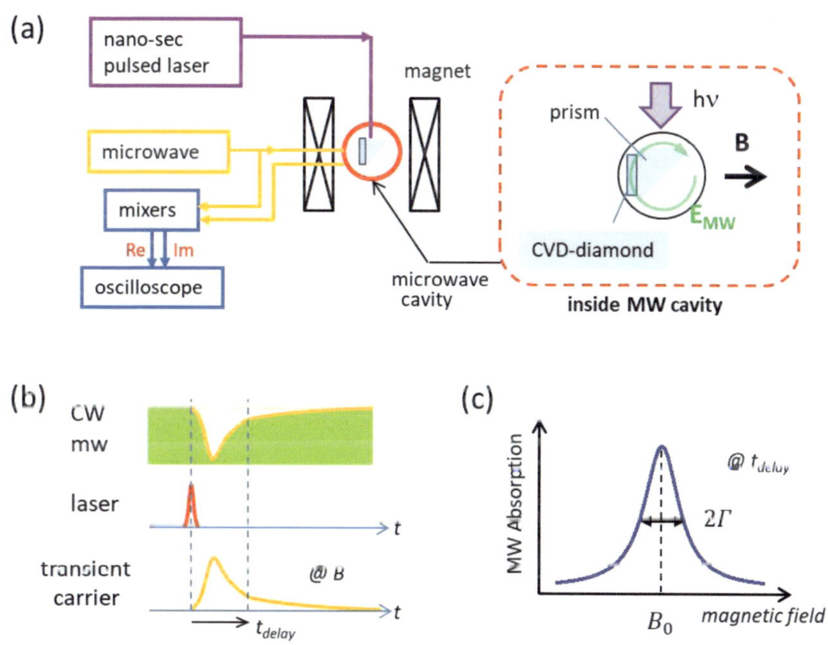

Figure 2.
Schematic drawings of experimental method: (a) the equipment and inside of a microwave cavity, (b) a temporal response of continuously applied microwave to transiently generated carrier by laser pulse at an external magnetic field, and (c) an example of a CR peak in spectrum at a delay time.

detection using microwave mixers and a two-channel oscilloscope of a system (Bruker, ELEXSYS E580) (**Figure 2b**). Inphase and out-of-phase signals to the input microwave were obtained as real and imaginary parts. The cavity's quality factor Q was set at less than 800, corresponding to a time resolution of less than 13 ns by a formula, $\Delta t = Q/\omega$. By sweeping the external magnetic field step-by-step, we obtained the set of the temporal response curves, from which a CR spectrum at a delay time was extracted as a function of the magnetic field as described in Section 3.1. Excitation spectra were also obtained by measuring the signal intensity at an external magnetic field as a function of the excitation wavelength at 10, 80, and 300 K. Although the CR spectrum was too broad to resolve any carrier species at higher temperatures, the amount of free carrier was estimated from the intensity of temporal response under the magnetic field. In addition, the signal intensity was measured according to the rise of temperature.

Important carrier parameters were extracted from a resonance peak in the CR spectrum (**Figure 2c**): the effective mass m^* from the resonance magnetic field B_0 by $m^* = eB_0/\omega = eB_0/2\pi\nu$, the carrier scattering time τ from the spectral half width at half maximum Γ by $1/\tau = \omega\Gamma/B_0$, and carrier drift mobility $\mu = e\tau/m^*$, where e is the elementary charge of electron. The carrier scattering time is also called as momentum relaxation time. For the analysis of effective masses, CR spectra were measured as a function of the magnetic field orientation, where the crystal was rotated about [1–10] crystalline axis so that the external magnetic field was oriented in the (1–10) plane or rotated about [100] axis with the external magnetic field in the (100) plane. Temperature dependence of CR spectral width provides the aspects of carrier scattering mechanism.

In addition to the abovementioned parameters, although we will not describe details here, important properties of carrier generation and decay can be unveiled from the time-resolved cyclotron resonance method: analysis of the rise time of the temporal curve and the signal intensity depending on excitation laser intensity can reveal a carrier generation mechanism [22, 29]. A lifetime of the carrier in a rotating motion is extracted from the decay time of a temporal curve. Temporal variation of carrier density is also estimated based on the plasma shift analysis [16, 30, 31]. Here, to study the basic properties of carriers, we paid careful attention to minimize plasma shifts of the resonance peaks, with the incident pulse energy less than 5.8 μJ which ensures the carrier density at the delay times later than 600 ns is less than 10^{11} cm^{-3}.

3. Results

3.1 Time-resolved cyclotron method

Figure 3a shows a colored contour map of a real part of TRCR signal measured at 7.3 K excited by laser pulses at photon energy of 5.50 eV. Temporal profiles at the magnetic fields of 0.089, 0.122, 0.162, and 0.230 T are shown in **Figure 3b**. CR spectra at the delay times of 60, 200, and 600 ns are shown in **Figure 3c**, by slicing the data set at the delay times. The magnetic field was applied to an angle of 40° from the crystal axis of [001] in the (1–10) plane. In this orientation, four carrier species, light hole, heavy hole, and two electrons in inequivalent conduction valleys, indicated by lh, hh, e1, and e2, respectively, were distinguishable as shown in **Figure 3a** and **c**.

DOI: http://dx.doi.org/10.5772/intechopen.86002

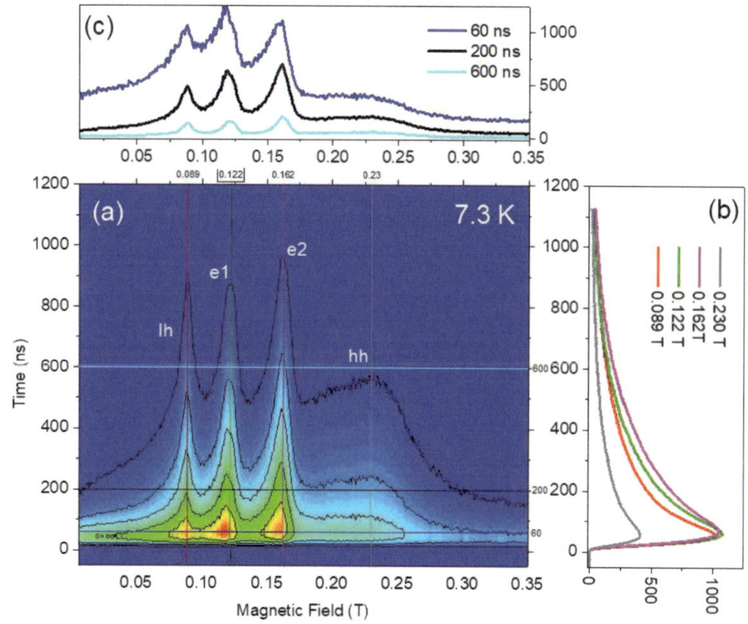

Figure 3.
(a) A colored contour map of a real part of TRCR signal at 7.3 K excited by laser pulses of photon energy at 5.50 eV; (b) temporal profiles at 0.089, 0.122, 0.162, and 0.230 T denoted by vertical lines in (a); (c) CR spectra at the delay times of 60, 200, and 600 ns denoted by horizontal lines in (a).

3.2 Optical carrier injection

Optical carrier injection is a key technique in our nanosecond TRCR method. As a diamond has an indirect band structure as shown in **Figure 1a**, the optical carrier injection at the lowest photon energy is established with the assistance of phonon emission/absorption to satisfy the energy and momentum conservations. The lowest excited state is an exciton band located below the indirect band edge by a binding energy larger than 80 meV [32, 33], whose fine structures were recently clarified [34]. To clarify the spectroscopic way of carrier injection, an excitation spectrum of TRCR signal at the fixed resonant magnetic field was measured with a thin CVD crystal of 70-μm in thickness to suppress the saturation by exciton absorption.

Figure 4 shows the TRCR excitation spectra obtained at 10, 80, and 300 K. The signals were averaged at the time windows, (a) 80–280 ns at 10 K, (b) 352–552 ns at 80 K, and (c) 156–356 ns at 300 K, after the signal decayed to the 1/e of the peak intensity [26]. For such late times, we observed that carriers were dominantly generated by dissociation of excitons [27]. The onset energy of the excitation spectra (a, b) at 5.493 eV coincides with the exciton generation edge assisted by emission of a transverse acoustic (TA) phonon ($E_{ex} + \hbar\omega_{TA}$), where E_{ex} = 5.406 eV is the exciton energy and $\hbar\omega_{TA}$ = 87 meV is the TA phonon energy [35] (see the inset). The second onset at 5.547 eV is assigned to exciton generation assisted by emission of a transverse optical (TO) phonon ($E_{ex} + \hbar\omega_{TO}$), where $\hbar\omega_{TO}$ = 141 meV is the TO phonon energy [35]. Therefore, in the range of the excitation photon energy above 5.493 eV, free carriers can be generated via excitons that are generated by the assistance of phonon emission. At the lower temperatures as 10 K, free carriers were generated by two-body collision of excitons, as the CR signal intensity was proportional to the square of excitation intensity [22].

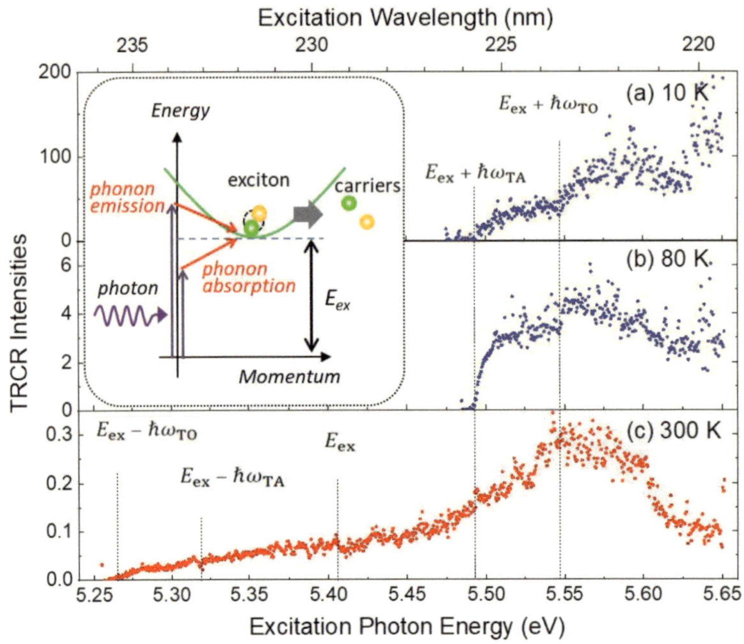

Figure 4.
CR excitation spectra at (a) 10 K, (b) 80 K, and (c) 300 K. Vertical broken lines indicate the energy positions of $E_{ex}\pm\hbar\omega_{TA}$, $E_{ex}\pm\hbar\omega_{TO}$, and E_{ex}. The inset shows schematic of phonon-assisted transitions to the exciton band. This figure was taken from [26] with a slight modification.

On the other hand, the signal at 300 K arose at the lower energy side with the onset at 5.265 eV. The onset energy coincides with the threshold for exciton generation assisted by absorption of a TO phonon ($E_{ex} - \hbar\omega_{TO}$) (see the inset). The onset energy for TA phonon absorption ($E_{ex} - \hbar\omega_{TA}$) is higher as indicated by the vertical broken line (the second one from the left) in **Figure 4c**. In the range of the excitation photon energy from 5.265 to 5.493 eV, free carriers can be generated only at higher temperatures via excitons that are generated by the assistance of phonon absorption. As the CR signal intensity measured at 300 K was proportional to the excitation intensity, the carriers are generated dominantly via a one-body process, such as thermal ionization of excitons [26].

Under the excitation in the range from 5.265 to 5.493 eV, where only the phonon absorption assists the process, the carrier number should increase with rising of the temperature according to the activation of phonons. The lower panel of **Figure 5** shows temperature dependence of the temporal response intensity excited by laser pulse at 5.335 eV. Solid curves in the upper panel are the temperature dependence of the quantum statistical numbers $\langle n \rangle = 1/\left(\exp\left(\hbar\omega_i/k_B T\right) - 1\right)$ of TA and TO phonons, where $\hbar\omega_i$, k_B, and T are the phonon energies for $i = TA$ or TO, Boltzmann constant, and temperature, respectively. The observed signal intensity rose around 150 K in coincidence to the appearance of TA phonon. This suggests that the phonon-assisted optical carrier injection is effective at device-operating temperatures.

In the subsequent Sections 3.3–3.5, we focus on the carrier properties at temperatures below 50 K. This temperature range is uniquely reached by our method owing to the optical carrier injection without the need of thermal activation of carriers from deep levels. For an efficient carrier generation at these temperatures, the excitation wavelength was chosen in the range of 219.4–226.4 nm. Furthermore, we discuss the CR spectra at the later delay times after 600 ns (see **Figure 3**) by eliminating the plasma shift effect at the earlier delay times depending on experimental conditions [16, 31].

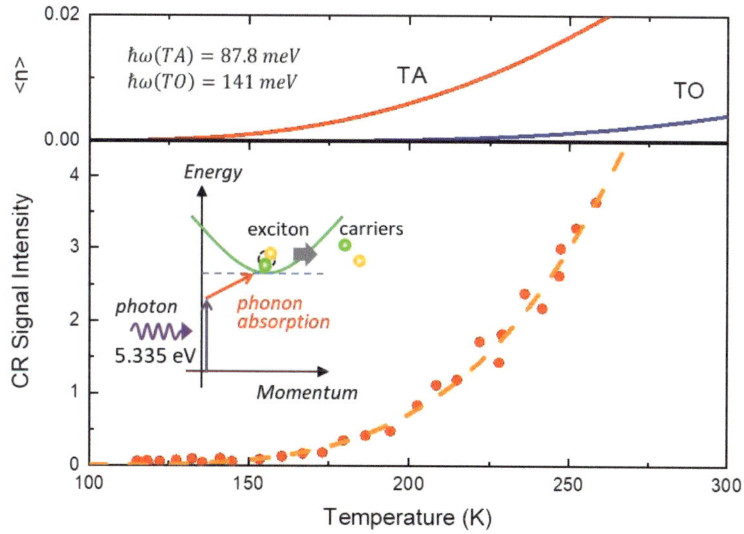

Figure 5.
Temperature dependence of the temporal response signal intensity at 0.16 mT due to free carriers generated by laser pulse at 5.335 eV via phonon absorption. A broken curve in the lower panel is a guide to eyes. Solid curves in the upper panel show the temperature dependence of the quantum statistical number ⟨n⟩ of TA or TO phonon.

3.3 Determination of effective masses

Both CR spectra of real and imaginary parts were well fitted by the formula for the complex conductivity [15]:

$$S(B) = \sum_j \frac{1 - iB_j/\Gamma_j}{\{(1 - iB_j/\Gamma_j)^2 + B^2/\Gamma_j^2\}}, \tag{1}$$

where Γ_j is a half width and B_j is a central magnetic field at the resonance taken to be common for both real and imaginary parts, B is an external magnetic field, and j represents each resonant component ($j = lh, hh, e1, e2$). We can obtain the effective mass m_j^* and momentum relaxation time τ_j from these fitting parameters by using the relations, $m^* = eB_0/\omega$ and $1/\tau_j = \omega\Gamma_j/B_j$. **Figure 6** plots the extracted effective masses m_j^* at the magnetic field orientations: the positive angles indicate the magnetic field in the (1–10) plane for the rotation axis along [1–10], whereas the negative angles indicate the magnetic field in the (100) plane for the rotation axis along [100].

The effective masses of electrons were simulated according to the following equation [15]:

$$m_{ei}^* = \sqrt{\frac{m_t^2 m_l}{m_t(1 - g_i(\theta)^2) + m_l g_i(\theta)^2}}, i = 1, 2, 3, \tag{2}$$

where $g_1(\theta) = \pm sin\theta/\sqrt{2}$, $g_2(\theta) = g_3(\theta) = \pm cos\theta$ on (1–10) plane and $g_1(\theta) = \pm sin\theta, g_2(\theta) = \pm cos\theta, g_3(\theta) = 0$ on (100) plane. The curves calculated with parameters $m_t/m_0 = 0.280$ and $m_l/m_0 = 1.560$ [21] trace the data points very well. The curves of e1 and e2 in the (1–10) plane originate from two and four equivalent valleys, respectively. The curve of e2 splits into two curves in the (100) plane, where electron of a constant effective mass is referred as e3.

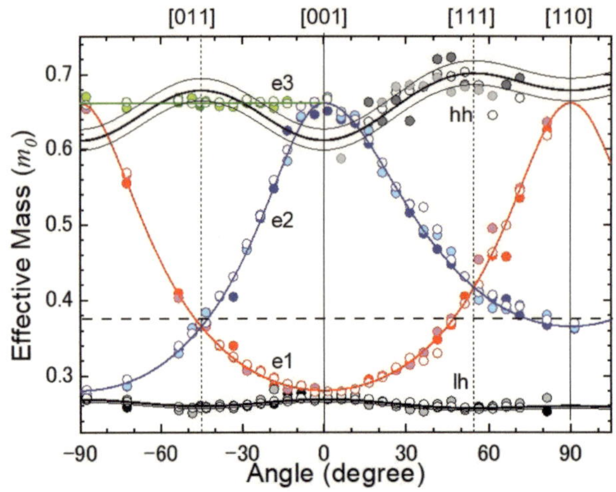

Figure 6.
Angular dependence of effective masses, taken from Ref. [21].

On the other hand, the effective masses of holes were simulated according to the equation for light (−) and heavy (+) holes [15]:

$$m_h^* \sim -\frac{1}{A \pm B'}\left\{1 \pm \frac{1}{A \pm B'}\frac{C^2}{B'}\frac{1}{16}f(\theta)\right\}, \tag{3}$$

where $f(\theta) = \left(1 - 3\cos^2\theta\right)^2/4$ in the (1–10) plane and $f(\theta) = \left(1 - 3\cos^2\theta\sin^2\theta\right)$ in the (100) plane. The thick lines for lh and hh were calculated with parameters $A = -2.670, B' = 1.245, C = 1.898$ [21]. The thin lines are calculated with the upper and lower limit of errors, $\Delta A = \pm 0.020, \Delta B' = \pm 0.025, \Delta C = \pm 0.048$. Moreover, the split-off hole mass calculated from the parameter as $m_{so}/m_0 \approx \hbar^2/(2m_0A)$ is also plotted by a broken red line. These parameters give band dispersion at the Γ point,

$$E(\mathbf{k}) = Ak^2 \pm \sqrt{B^2k^4 + C^2\left(k_x^2k_y^2 + k_y^2k_z^2 + k_z^2k_x^2\right)} \tag{4}$$

with a transformation by $B' = \sqrt{B^2 + C^2/4}$. The finite value of C gives rise to the warping of the valence band.

As we report in detail in Ref. [21], it is experimentally figured out that the electrons are in highly asymmetric valleys along the <001> directions, that is, at the $\triangle$ points, with the transverse effective mass ($m_t = 0.280\ m_0$) and longitudinal effective mass ($m_l = 1.560\ m_0$). And, the doubly degenerate valence-band maxima are located at the Γ point and warping. The values of effective masses were listed in **Table 1** in comparison to those of silicon and germanium [15]. The calculated values of the density of mass for electron $m_{ed} = \left(m_t^2 m_l\right)^{1/3}$ and hole $m_{hd} = \left(m_{lh}^{3/2} + m_{hh}^{3/2} + m_{SO}^{3/2}\right)^{2/3}$ were also listed with the value of direct bandgap energy at Γ point.

Figure 7 compares the angular dependence of effective masses of diamond with those of silicon and germanium [15] with the same angular definition as in **Figure 6**. The conduction-band minimum in silicon is located at the $\triangle$ points as in a diamond, while that in germanium is located at the L points. It is easily recognized that the effective masses in diamond are largest among group-IV semiconductors, reflecting the largest direct bandgap energy at the Γ point which causes relatively larger contribution of the first perturbation term in the **k·p** theory to the effective mass of an energy band.

	E_g^{direct}	m_t	m_l	m_{lh}	m_{hh}	m_{SO}	m_{ed}	m_{hd}
	(eV)				(m_0)			
Ge[a]	0.80	0.082	1.58	0.043	0.336	0.095	0.220	0.378
Si[a]	3.4	0.19	0.97	0.16	0.52	0.23	0.327	0.670
Diamond	7.3	0.280	1.56	0.260	0.667	0.375	0.496	0.947

[a]*Values of germanium and silicon were taken from Ref. [15].*

Table 1.
Comparison of effective mass in group-IV semiconductor.

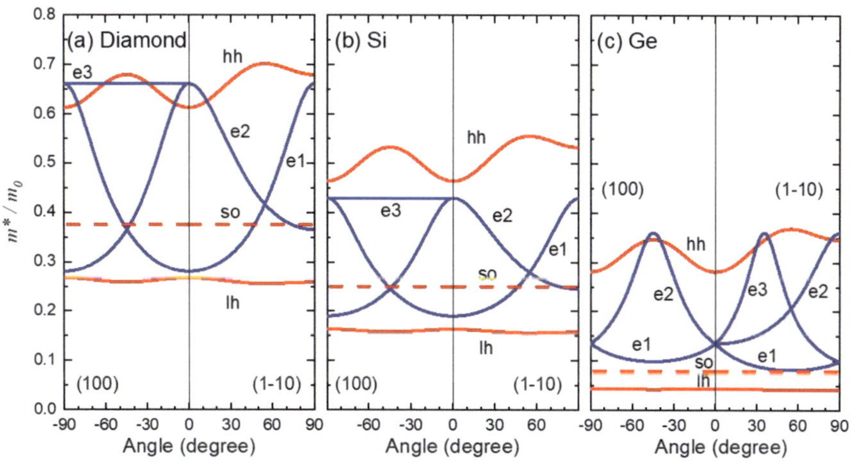

Figure 7.
Comparison of effective mass in group-IV semiconductor: angular dependence of effective masses simulated with the experimentally obtained parameters, in (a) diamond [21], (b) silicon [15], and (c) germanium [15]. Three electrons (blue lines) and light and heavy holes (red lines) are plotted for two rotation planes; the negative angles mean the magnetic field in the (100) plane for the rotation axis along [100], whereas the positive angles mean the magnetic field in the (1–10) plane for the rotation axis along [1–10].

3.4 Sample dependence of carrier lifetime

A well-resolved spectrum of TRCR at the lower temperatures allows extracting the effective masses in good accuracy as described in Section 3.3. We compared the TRCR signals in different samples as reported in Refs. [24, 25]. The sample showed the narrower spectral width as presented in **Figure 3** possesses the smaller concentration of donor and acceptor, that is, nitrogen and boron. **Figure 8** shows temporal profiles of five different samples. The sample displayed a slow rise and decay in a couple of hundred nanoseconds (sample A) which is identical to that in **Figure 3**. The narrow spectral width is caused by the long carrier scattering time. From the comparison of CR spectra of CVD diamonds to those of dislocation-free HPHT diamonds, we found the fact that the TRCR detection is rather insensitive to crystalline dislocations [21]. It is known that a typical dislocation density in CVD diamond is lower than 10^4 cm^{-2} [36], corresponding to dislocation periods larger than 100 μm. On the other hand, the cyclotron radii in the measurement with X band microwave were 86 and 55 nm for light and heavy holes, respectively. As the carriers rotate in the much smaller spatial extension than the typical dislocation period in CVD samples, the CR detection is rather insensitive to dislocations. Instead, impurity scattering by neutral nitrogen atoms is found to be dominant at low temperatures (as described in Section 3.5), because their average separations are comparable to the cyclotron radii in the present case.

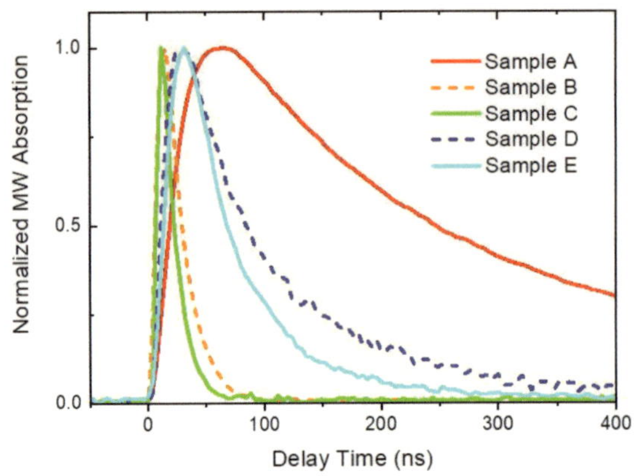

Figure 8.
Temporal profiles at the resonance of light hole in different five samples listed in ***Table 2****, taken from Ref. [25].*

From these facts, we emphasize that the accurate determination of effective masses as described in Section 3.3 became possible, since we could use a highly pure diamond produced by the CVD method under the optical carrier injection.

The rise time and decay time in the temporal profile of TRCR in **Figures 3b** and **8** reveal carrier generation and trapping mechanism. The finite rise time reflects the time required for carrier creation by exciton collision. A detailed formula giving an approximate rise time in connection with the lifetime is described in Refs. [22, 29]. The shorter decay time is probably caused by the higher density of impurity concentrations (in comparison among samples A–C) and by the higher density of stacking faults and substitutional impurities (in comparison samples E–D). It has been known that incorporation of defects occurs more easily in a (111)-oriented diamond than in an (001)-oriented diamond.

3.5 Carrier scattering time and drift mobility

The temperature dependence of TRCR spectrum provides the aspect of carrier scattering mechanisms. **Figure 9a** shows the normalized temporal curves measured at 0.16 mT at various temperatures. The rise time and the decay time of the signal increased as the temperature is rising. The longer rise and decay times at higher temperatures indicate elongating of the carrier lifetime [22]. This is probably caused by trapping of carriers into impurity states at lower temperatures. Similar shortening of the exciton lifetime at low temperatures was clarified in Ref. [37] by comparing exciton lifetimes in samples containing different concentrations of impurities.

Sample	Growth	Boron	Nitrogen	Dislocation
A	CVD (001)-sector	<1 ppb	<5 ppb	–
B	CVD (001)-sector	<50 ppb	<100 ppb	–
C	HPHT+neutron irrad.	–	51 ppm	–
D	HPHT (001)-sector	<0.8 ppb	<45 ppb	Free
E	HPHT (111)-sector	<0.8 ppb	<45 ppb	Free

"-" means unknown.

Table 2.
Specification of used samples.

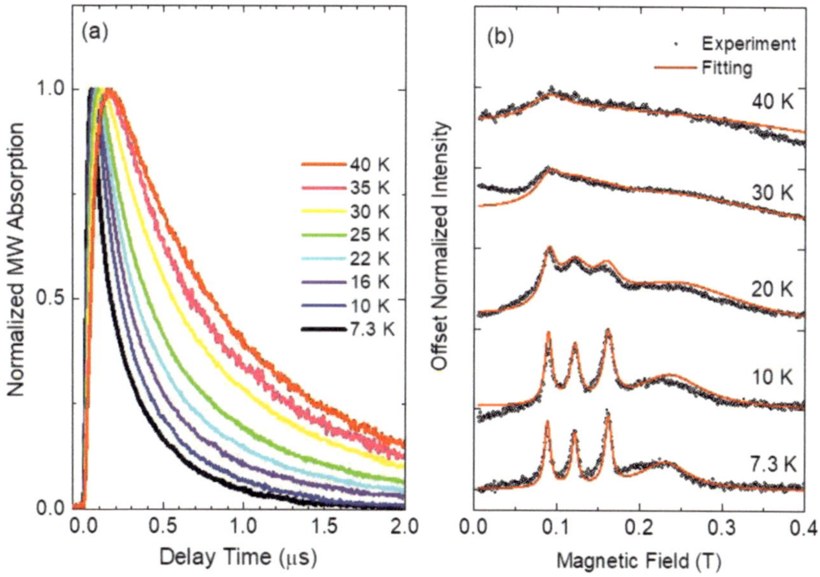

Figure 9.
Temperature variation of temporal curves at 0.16 mT (a) and spectrum at the delay time 1 μs averaged for ±40 ns (b), adapted from Ref. [23].

Figure 9b shows the CR spectra at 7.3, 10, 20, 30, and 40 K taken at a delay time of 1 μs after the laser pulse with averaging window for ±40 ns [23]. The four peaks were separately observed up to 20 K and the width broadened with increasing temperature. These spectra were analyzed by the abovementioned spectrum fitting. The carrier scattering times τ_j's, in other words, momentum relaxation times, were taken from the width Γ_j as a function of temperature. The fitting of the structureless spectra at the higher temperatures was possible by fixing the resonance field of the four components. The fitting curve of each spectrum is shown by the thin red lines.

The spectral width Γ_j at low temperature sharply depends on the impurity concentration as mentioned in Section 3.4. The carrier scattering at low temperature is mainly governed by the neutral impurity scattering instead of the ionized impurity, because the optically injected carriers neutralize ionized impurity centers. **Figure 10a** shows the temperature dependence of the inverse carrier scattering times $1/\tau_j$ extracted from the spectra measured with a CVD diamond shown in **Figure 9b**. The data points are well reproduced by the sum of the longitudinal acoustic (LA) phonon scattering rate $1/\tau_{acj}$ and a constant term b_j (solid lines). The LA phonon scattering rate $1/\tau_{acj}$ (broken lines) was calculated for electrons and holes by using Eq. (6) in Ref. [38] with a deformation potential of 8.7 eV [35] (6.8 eV) for e1 (e2) electrons or 10.0 eV for both types of holes in diamond. Deviation from the straight line of the $T^{3/2}$ law which was known for acoustic phonon scattering appeared below 40 K due to inhibition of LA phonon emission at low temperatures. The constant terms were attributed to neutral impurity scattering rates. The best fit values were $b_{e1} = 1.4 \times 10^9\ s^{-1}$, $b_{e2} = 1.3 \times 10^9\ s^{-1}$, $b_{lh} = 2 \times 10^9\ s^{-1}$, and $b_{hh} = 8 \times 10^9\ s^{-1}$. As discussed in Ref. [23], these values were larger than the values calculated by modified Erginsoy's formulas, $\beta_e = (3.4 n_A a_A + 20 n_D a_D)\hbar/m_e^*$ for the electron [39] and $\beta_h = (20 n_A a_A + 3.4 n_D a_D)\hbar/m_h^*$ for the hole, where $a_{A(D)}$ is a Bohr radius and $n_{A(D)}$ is the concentration of acceptors (donors). This fact indicates that the estimated $n_I a_I$ $(I = A, D)$ were smaller than those in the actual situations probably owing to the too much simplified estimation of the Bohr radius by the hydrogenic model.

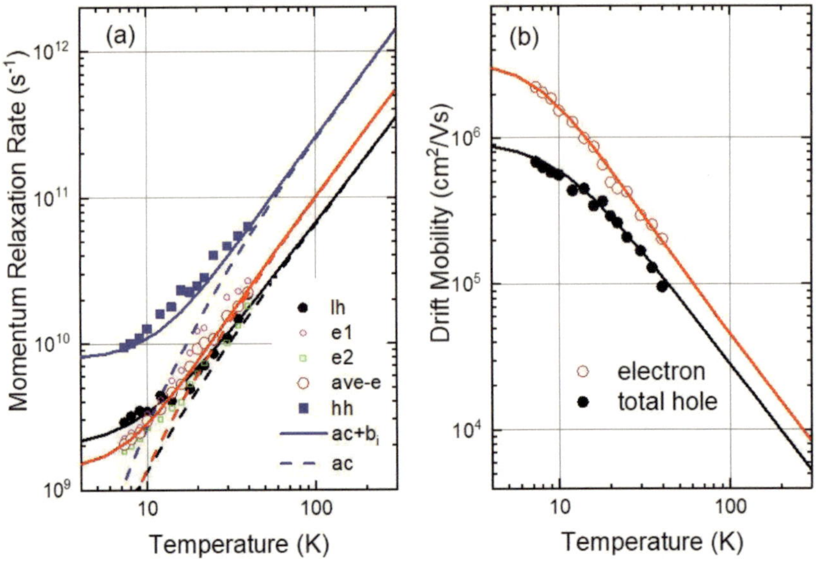

Figure 10.
Temperature dependences of carrier scattering time (a) and drift mobility (b), taken from Ref. [23].

	Mobility at 300 K (cm^2/Vs)	
	μ_e	μ_h
Ge	3900[a]	1900[a]
Si	1450[a]	500[a]
Diamond	7300[b]	5300[b]

[a]*Taken from Ref. [42].*
[b]*Present extrapolated values.*

Table 3.
Comparison of mobility at 300 K.

Now as the parameters of the effective mass m^* and carrier scattering time τ were individually obtained from the analysis of CR spectra, we can derive the drift mobility by the relation $\mu = e\tau/m^*$. **Figure 10b** shows the drift mobilities of electron and hole. We used a conductivity mass for the electron, $m_e^* = 3/(1/m_l + 2/m_t) = (0.386 \pm 0.026)\,m_0$. For the valence bands, effective masses were averaged for the warped energy surfaces $m_{lh}^*/m_0 = 0.261 \pm 0.005$ for the light hole and $m_{hh}^*/m_0 = 0.663 \pm 0.032$ for the heavy hole. The total hole mobility was obtained by weighting the light and heavy hole mobilities as $\mu = (p_{lh}\mu_{lh} + p_{hh}\mu_{hh})/(p_{lh} + p_{hh}) = (\mu_{lh} + r\mu_{hh})/(1 + r)$ with the ratio $r = p_{hh}/p_{lh} = (m_{hh}/m_{lh})^{3/2} = 4.05$ derived from the density-of-state ratio for the degenerate energy bands [40]. The drift mobility at 10 K was obtained as $\mu_e = (1.52 \pm 0.27) \times 10^6\,cm^2/Vs$ for electrons and $\mu_{lh} = (2.26 \pm 0.38) \times 10^6$ and $\mu_{hh} = (0.24 \pm 0.03) \times 10^6\,cm^2/Vs$ for the light and heavy holes. The ratio of light to heavy hole mobilities $\mu_{lh}/\mu_{hh} \cong 9.35$ is in better agreement with the ratio of $(m_{hh}^*/m_{lh}^*)^{5/2} = 10.3$ than $m_{hh}^*/m_{lh}^* = 2.54$. This fact indicates that light and heavy holes in ultrapure diamond relax inside each band. This is in contrast to the case in p-type germanium [40], where the intra-/inter-band scattering is dominant for the heavy-/light-hole and then μ_{lh}/μ_{hh} is proportional to m_{hh}^*/m_{lh}^*.

We evaluated the mobility up to 300 K by extrapolating the $T^{3/2}$ relation as shown by the solid lines, because the optical phonon process dominates carrier scattering only above 400 K in diamond [41]. The values of the mobility at 300 K were derived as $\mu_e = 7.3 \times 10^3\,\mathrm{cm^2/Vs}$ for electron and $5.3 \times 10^3\,\mathrm{cm^2/Vs}$ for total hole. The values are higher than those evaluated by Reggiani et al. [8] and Nava et al. [9] for a natural n(p)-type diamond crystal or by Isberg et al. [10, 17] with the CVD diamond crystals. The values are summarized in **Table 3** in comparison to those in silicon and germanium at 300 K [42]. The highest mobility of both electron and hole responsive to 10 GHz at 300 K indicates the outstanding possibility of a diamond in the field of power electronics and optoelectronics.

4. Conclusion

Recently developed experimental method, the nanosecond time-resolved cyclotron resonance, was introduced to clarify the basic carrier transport parameters in an intrinsic diamond. A sophisticated optical carrier injection technique in a highly pure diamond crystal realized the measurement at low temperature. The extracted effective masses, carrier scattering times, and mobilities unveiled the supreme carrier transport properties of a highly pure diamond, which indicate a large application-oriented advantage especially in power electronics and optoelectronics fields. The introduced optical carrier injection is a promising technique to control a carrier density in future devices.

Acknowledgements

The authors thank J.H. Kaneko (Hokkaido University) for providing the diamond sample grown by the CVD method and Ms. S. Hamabata (Wakayama University) for the experiments described in Section 3.2. This work was supported by JSPS KAKENHI (Grant Nos. 15 K05129 and 17H02910) and the Murata Science Foundation.

Author details

Ikuko Akimoto[1]* and Nobuko Naka[2]

1 Wakayama University, Wakayama, Japan

2 Kyoto University, Kyoto, Japan

*Address all correspondence to: akimoto@sys.wakayama-u.ac.jp

References

[1] Hiraiwa A, Kawarada H. Figure of merit of diamond power devices based on accurately estimated impact ionization processes. Journal of Applied Physics. 2013;**114**:034506. DOI: 10.1063/1.4816312

[2] Isberg J, Gabrysch M, Hammersberg J, Majdi S, Kovi KK, Twitchen DJ. Generation, transport and detection of valley-polarized electrons in diamond. Nature Materials. 2013;**12**:760. DOI: 10.1038/NMAT3694

[3] Makino T, Yoshino K, Sakai N, Uchida K, Koizumi S, Kato H, et al. Enhancement in emission efficiency of diamond deep-ultraviolet light emitting diode. Applied Physics Letters. 2011;**99**:061110. DOI: 10.1063/1.3625943

[4] Breeze JD, Salvadori E, Sathian J, Alford NM, Kay CWM. Continuous-wave room-temperature diamond maser. Nature. 2018;**555**:493. DOI: 10.1038/nature25970

[5] Kosaka H, Niikura N. Entangled absorption of a single photon with a single spin in diamond. Physical Review Letters. 2015;**114**:053603. DOI: 10.1103/PhysRevLett.114.053603

[6] Weber JR, Koehl WF, Varley JB, Janotti A, Buckley BB, Van de Walle CG, et al. Quantum computing with defects. PNAS. 2010;**107**:8513. DOI: 10.1073/pnas.1003052107

[7] Balmer RS, Brandon JR, Clewes SL, Dhillon HK, Dodson JM, Friel I, et al. Chemical vapour deposition synthetic diamond: Materials, technology and applications. Journal of Physics. Condensed Matter. 2009;**21**:364221. DOI: 10.1088/0953-8984/21/36/364221

[8] Reggiani L, Bori S, Canali C, Nava F, Kozlov SF. On the lattice scattering and effective mass of holes in natural diamond. Solid State Communications. 1979;**30**:333. DOI: 10.1016/0038-1098(79)90645-8

[9] Nava F, Canali C, Jacobini C, Reffiani L, Kozlov SF. Electron effective masses and lattice scattering in natural diamond. Solid State Communications. 1980;**33**:475. DOI: 10.1016/0038-1098(80)90447-0

[10] Isberg J, Hammersberg J, Johansson E, Wilstroem T, Twitchen DJ, Whitehead AJ, et al. High carrier mobility in single-crystal plasma-deposited diamond. Science. 2002;**297**:1670. DOI: 10.1126/science.1074374

[11] Kono J, Takeyama S, Takamasu T, Miura N, Fujimori N, Nishibayashi Y, et al. High-field cyclotron resonance and valence-band structure in semiconducting diamond. Physical Review B. 1993;**48**:10917. DOI: 10.1103/PhysRevB.48.10917

[12] Rauch CJ. Millimeter cyclotron resonance experiment in diamond. Physical Review Letters. 1961;7:83. DOI: 10.1103/PhysRevLett.7.83

[13] Redfield AG. Electronic hall effect in diamond. Physics Review. 1954;**94**:526. DOI: 10.1103/PhysRev.94.526

[14] Denham P, Lightowlers EC, Dean PJ. Ultraviolet intrinsic and extrinsic photoconductivity of natural diamond. Physics Review. 1967;**161**:762. DOI: 10.1103/PhysRev.161.762

[15] Dresselhaus G, Kip AF, Kittel C. Cyclotron resonance of electrons and holes in silicon and germanium crystals. Physics Review. 1955;**98**:368. DOI: 10.1103/PhysRev.98.368

[16] Dresselhaus G, Kip AF, Kittel C. Plasma resonance in crystals: Observations and theory. Physics Review. 1955;**100**:618. DOI: 10.1103/PhysRev.100.618

[17] Gabrysch M, Majdi S, Twitchen DJ, Isberg J. Electron and hole drift velocity in chemical vapor deposition diamond. Journal of Applied Physics. 2011;**109**:063719. DOI: 10.1063/1.3554721

[18] Majdi S, Kovi KK, Hammersberg J, Isberg J. Hole transport in single crystal synthetic diamond at low temperatures. Applied Physics Letters. 2013;**102**:152113. DOI: 10.1063/1.4802449

[19] Ščajev P, Gudelis V, Ivakin E, Jarašiūnas K. Nonequilibrium carrier dynamics in bulk HPHT diamond at two-photon carrier generation. Physica Status Solidi A: Applications and Materials Science. 2011;**208**:2067. DOI: 10.1002/pssa.201100006

[20] Nesladek M, Bogdan A, Deferme W, Tranchant N, Bergonzo P. Charge transport in high mobility single crystal diamond. Diamond and Related Materials. 2008;**17**:1235. DOI: 10.1016/j.diamond.2008.03.015

[21] Naka N, Fukai K, Handa Y, Akimoto I. Direct measurement via cyclotron resonance of the carrier effective masses in pristine diamond. Physical Review B. 2013;**88**:035205. DOI: 10.1103/PhysRevB.88.035205

[22] Naka N, Fukai K, Handa Y, Akimoto I. Nanosecond cyclotron resonance in ultrapure diamond. Journal of Luminescence. 2014;**152**:93. DOI: 10.1016/j.jlumin.2013.12.055

[23] Akimoto I, Handa Y, Fukai K, Naka N. High carrier mobility in ultrapure diamond measured by time-resolved cyclotron resonance. Applied Physics Letters. 2014;**105**:032102. DOI: 10.1063/1.4891039

[24] Akimoto I, Naka N, Tokuda N. Time-resolved cyclotron resonance on dislocation-free HPHT diamond. Diamond and Related Materials. 2016;**63**:38. DOI: 10.1016/j.diamond.2015.08.013

[25] Naka N, Morimoto H, Akimoto I. Excitons and fundamental transport properties of diamond under photo-injection. Physica Status Solidi A: Applications and Materials Science. 2016;**213**:2551. DOI: 10.1002/pssa.201600237

[26] Hamabata S, Akimoto I, Naka N. Temperature-dependent carrier injection routes under optical excitation in high-purity diamond crystals. Journal of Physics: Conference Series; accepted

[27] Akimoto I, Hamabata S, Kaneko JH, Naka N. in preparation

[28] Akimoto I, Naka N. Two optical routes of cold carrier injection in silicon revealed by time-resolved excitation spectroscopy. Applied Physics Express. 2017;**10**:061301. DOI: 10.7567/APEX.10.061301

[29] Naka N, Akimoto I, Shirai M. Free-carrier generation by two-photon resonant excitation to the excitonic states in cuprous oxide. Physica Status Solidi B. 2013;**250**:1773. DOI: 10.1002/pssb.201200713

[30] Naka N, Akimoto I, Shirai M, Kan'no K. Time-resolved cyclotron resonance in cuprous oxide. Physical Review B. 2012;**85**:035209. DOI: 10.1103/PhysRevB.85.035209

[31] Akimoto I, Torai S, Naka N, Shirai M. Temporal shift from magnetoplasma resonance to cyclotron resonance of photo carriers generated from 1 s-exciton on cuprous oxide crystal. European Physical Journal B. 2012;**85**:374. DOI: 10.1140/epjb/e2012-30618-8

[32] Clark CD, Dean PJ, Harris PV. Intrinsic edge absorption in diamond. Proceedings of the Royal Society A. 1964;**277**:312. DOI: 10.1098/rspa.1964.0025

[33] Dean PJ, Male JC. Luminescence excitation spectra and recombination

radiation of diamond in the fundamental absorption region. Proceedings of the Royal Society A. 1964;**277**:330. DOI: 10.1098/rspa.1964.0026

[34] Hazama Y, Naka N, Stolz H. Mass-anisotropy splitting of indirect excitons in diamond. Physical Review B. 2014;**90**:045209. DOI: 10.1103/PhysRevB.90.045209

[35] Jacoboni C, Reggiani L. The Monte Carlo method for the solution of charge transport in semiconductors with applications to covalent materials. Reviews of Modern Physics. 1983;**55**:645. DOI: 10.1103/RevModPhys.55.645

[36] Martineau PM, Gaukroger MP, Guy KB, Lawson SC, Twitchen DJ, Friel I, et al. High crystalline quality single crystal chemical vapour deposition diamond. Journal of Physics. Condensed Matter. 2009;**21**:364205. DOI: 10.1088/0953-8984/21/36/364205

[37] Naka N, Omachi J, Sumiya H, Tamasaku K, Ishikawa T, Kuwata-Gonokami M. Density-dependent exciton kinetics in synthetic diamond crystals. Physical Review B. 2009;**80**:035201. DOI: 10.1103/PhysRevB.80.035201

[38] Trauernicht DP, Wolfe JP. Drift and diffusion of paraexcitons in Cu_2O: Deformation-potential scattering in the low-temperature regime. Physical Review B. 1986;**33**:8506. DOI: 10.1103/PhysRevB.33.8506

[39] Erginsoy C. Neutral impurity scattering in semiconductors. Physics Review. 1950;**79**:1013. DOI: 10.1103/PhysRev.79.1013

[40] Brown DM, Bray R. Analysis of lattice and ionized impurity scattering in p-type Germanium. Physics Review. 1962;**127**:1593. DOI: 10.1103/PhysRev.127.1593

[41] Pernot J, Volpe PN, Omnes F, Muret P, Mortet V, Haenen K, et al. Hall hole mobility in boron-doped homoepitaxial diamond. Physical Review B. 2010;**81**:205203. DOI: 10.1103/PhysRevB.81.205203

[42] Sze SM, Ng KK. Physics of Semiconductor Devices. 3rd ed. New York: Wiley-Interscience; 1981. ISBN-10: 0471143235

Chapter 6

Simulation of Diamond Surface Chemistry: Reactivity and Properties

Karin Larsson

Abstract

The diamond material possesses very attractive properties, such as superior electronic properties (when doped), in addition to a controllable surface termination. During the process of diamond synthesis, the resulting chemical properties will depend not only on the adsorbed species but also on the type of substitutional doping element. The combination of adsorbate and dopant will thus have the ability to influence both the chemical and electronic properties of a diamond surface. All resulting (and interesting) properties of doped and terminated diamond surfaces make it clear that these types of material modifications are very important for a variety of applications that are based on photoactivated chemical processes. Theoretical modeling has been shown to act as an important scientific tool in explaining and predicting experimental results. Simulation of the dependence of, e.g. surface termination and doping on diamond material properties, is expected to give important information about various surface electronic properties (like photo-induced surface electrochemistry).

Keywords: diamond, doping, theory, electronic properties

1. Introduction

The diamond material possesses very attractive properties, such as high transparency, high thermal conductivity at room temperature, radiation hardness, as well as an extreme mechanical hardness. It is also chemically inert. In addition, diamond also exhibits superior electronic properties (including high carrier mobility), large electrochemical potential window, low dielectric constant, controllable surface termination, and a high breakdown voltage [1]. However, the large band gap (5.4 eV) limits the usage of diamond for electronic devices. Hence, impurities have been introduced into the diamond lattice in order to eliminate this limitation. Impurities within a diamond lattice can dramatically change both the geometrical and electronic properties.

More generally, the properties of a surface (for solid material) depend on various factors, such as (i) number of binding atoms (i.e. for the atoms in the diamond surface plane), (ii) surface reconstructing, (iii) type and degree of chemisorption to the surface, and (iv) elemental doping within the upper surface region. The number of binding atoms to a surface atom (i) is always smaller than for a bulk atom, making it more reactive (since it has the driving force to become bulk-like).

This is also the underlying reason to the specific surface electronic properties, which differ from the corresponding bulk scenario. Moreover, factors (ii)–(iv) do all represent changes in the surrounding electron density for the surface atoms, of which all are different from the situations with bulk atoms.

The boron atom is one of the most commonly used p-type dopants. The B-doped diamond electrode is a semiconducting material with very promising properties like (i) a wider potential window in aqueous solution (approximately −1.35 to +2.3 V vs. the normal hydrogen electrode), (ii) low background current, and (iii) corrosion stability in aggressive environments [2]. Hence, boron-doped diamond (BDD) has been widely used for electronic applications. Nitrogen, as an n-type dopant, has also frequently been studied during the last decades [3–6]. However, the deep donor level of nitrogen (1.7 eV below the conduction band) limits the usage of nitrogen-doped diamond as an effective n-type semiconductor. Thus, phosphorus has instead been introduced as a plausible n-type donor in diamond (111) [7]. Phosphorus, P, is nowadays the only well-established substitutional n-type donor [8]. Kato et al. produced this type of n-type diamond epilayer on (100)-oriented diamond surfaces [9]. N-type diamond heavily doped with P atoms has more recently been obtained with low electric resistivity and low incorporation activation energy [10, 11]. This makes phosphorus-doped diamond more promising for device applications, e.g. as an electron source for bipolar devices [12–14], as Schottky barrier diodes [15], and for thermionic emission applications [16].

Surface termination is the general notation when a surface-binding species is used to uphold the cubic structure or to change the surface properties of diamond. For instance, hydrogen-terminated B-doped diamond surfaces have been found to be hydrophobic [17]. An H-terminated P-doped diamond surface has also been found to show negative electron affinity (NEA) [18]. (NEA means that electrons in the conduction band of diamond easily emit from the surface.) Electron emission from the conduction band of H-terminated P-doped homoepitaxial diamond (111) has experimentally been observed [16, 19]. The oxygen atom is also one of the most commonly used diamond surface-terminating species. Oxygen-terminated diamond surfaces do generally show hydrophilic properties [20], as well as a positive electron affinity. Oxygen termination can appear as a surface-binding OH, O_{bridge}, C_{ontop}, or OOH groups. Different chemical routes are used to link functional groups to these O-group adsorbates. This procedure is usually performed in, e.g. electroanalytical applications [21]. Fluorine-termination will, compared to H-termination, result in an even more hydrophobic diamond surfaces. In addition, F-terminated diamond surfaces exhibit exceptional electrochemical properties, such as a lower electrocatalytic activity, a wide electrochemical potential window, and low background current [22, 23]. Nitrogen-terminated diamond surfaces are generally very chemically reactive and are thereby commonly used as an intermediate for further modification of the diamond surface (e.g. as a linker). For instance, a diamond surface that is terminated with amine (NH_2) can be modified to be used as a functionalized surface in biosensors [24].

All these interesting properties of terminated diamond surfaces make it clear that surface termination is crucial for especially those applications for which diamond can function as an electrode material. The relative stability and the electronic structures of the variously doped diamond surfaces, terminated with either H, O-containing groups, F or NH_2, are therefore very important to study from an atomic-level point of view.

Theoretical modeling, using density functional theory (DFT), has been shown to be a valuable tool in the explanation and prediction of experimental results. For instance, DFT calculations have been shown powerful in studying thin film growth mechanisms, as well as for the simulation of surface reconstruction, modification, and functionalization, respectively.

2. Methods and methodologies

Theoretical modeling has been found useful in the (i) interpretation of experimental results and in the (ii) prediction of parameters in the design of experimental setups. First-principle DFT calculations have thereby become useful for a variety of materials of practical importance. This method is, however, quite demanding since it relies on the electrons in the systems under investigations. Hence, this circumstance puts a strong demand in the model size. It should here be stressed that these more advanced computational methods are needed when studying chemical reactions, where bonds are formed and/or broken. Other observables that are very sensitive to the theoretical method used, with its parameters, are the electronic structures with their intrinsic band gaps. In conclusion, first-principle DFT methods are very useful for studies of, e.g. electronic and geometrical surface structures, in addition to surface reactivity and chemical processes.

Periodic boundary conditions are most frequently used for DFT calculations on solid surfaces or interfaces. For this purpose, an ultrasoft pseudopotential [25] plane-wave approach is most often used, which, in turn, is based on the Perdew-Wang (PW91) generalized gradient approximation (GGA) [26] for the exchange-correlation functional. The GGA method takes into account the gradient of the electron density, which improves the accuracy of, e.g. the bond energies [27].

The model used under periodic boundary conditions are the so-called super cells where these cells will become translated in x-, y-, and z-directions, forming an infinite three-dimensional model. In the case of modeling surfaces, only a part of the supercell will be filled (i.e. in x- and y-directions), forming a repetitive pattern of infinite slabs with a vacuum layer in between. This vacuum layer has to be large enough to avoid any type of interactions between neighboring slabs.

3. Surface reactivities

There are two factors that are very important for a solid surface reactivity. One is the number of binding atoms for a surface atom. The other factor is the density of surface atoms. Bulk diamond C atoms bind covalently, with sp^3 hybridized bonds, to four C neighbors in. For the most commonly observed diamond facets (111) and (100), the (111) surface atoms only have three binding neighbors, whilst the corresponding (100) atoms have two (**Figure 1a**, **c**). (These surfaces are then assumed to be non-reconstructed.) When instead describing the situation in terms of dangling bonds (i.e. unpaired electrons), a (111) surface atom has only one dangling bond. On the other hand, the (100) surface has two dangling bonds, which makes it more reactive (as compared with the (111) surface). However, the presence of dangling bonds is most often the cause of a high surface reactivity. Hence, surface reconstructions may occur, with a resulting lowering of the surface energy. This is the reason why an ideal diamond (100) surface (i.e. (100)-1 × 1) most often undertakes a surface reconstruction, thereby forming a (100)-2 × 1 surface structure (**Figure 1d**). As a matter of fact, also the more ideal (111) surface (i.e. (111)-1 × 1) will at higher temperatures be transferred to a 2 × 1 reconstruction (i.e. (111)-2 × 1), called the Pandey chain structure (**Figure 1b**).

The relative energies for these surface structures have earlier been calculated theoretically by using an ab initio DFT method [28]. It was then shown that the 2 × 1-reconstruction of the bare diamond (100) surface will result in a stabilization of the total energy by 1.85 eV per surface C. The corresponding energy stabilization for the bare diamond (111) surface was found to be 0.54 eV per surface C

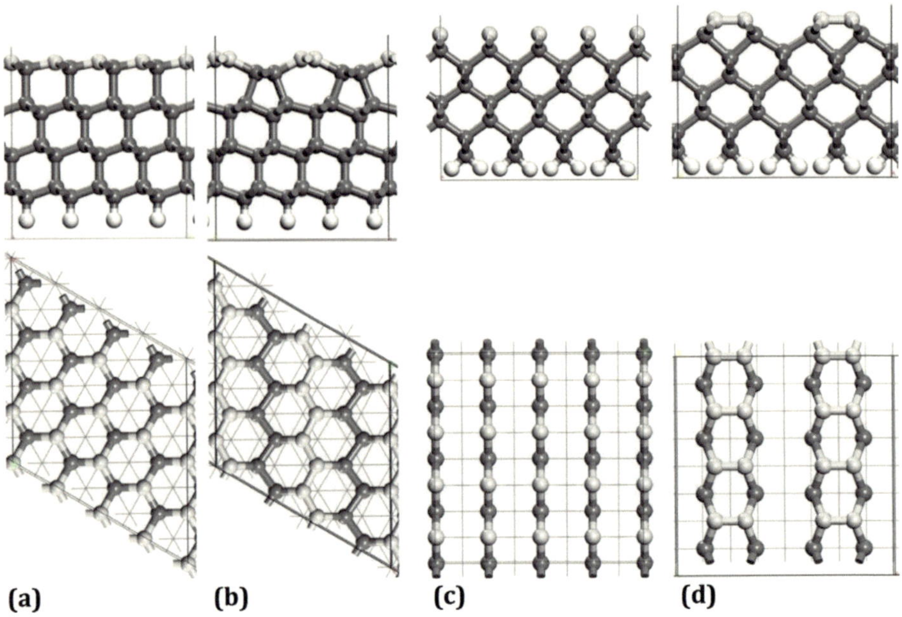

Figure 1.
Optimized periodic models of diamond (111)-1 × 1 (a), (111)-2 × 1 (b), (100)-1 × 1 (c), and (100)-2 × 1 (d) surfaces. The upper row shows a side view of the different models, and the lower row shows on-top orientations. The top-most C atoms are shown with a whiter shade.

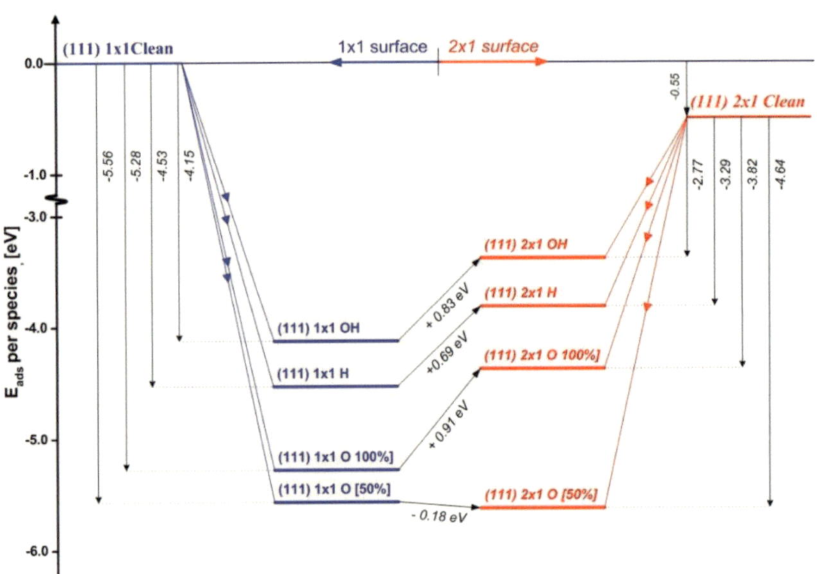

Figure 2.
Adsorption energies for various terminating species (H, OH, and O) onto (1 × 1)- and (2 × 1)-reconstructed diamond (100) and (111) surfaces.

(see **Figure 2**). These stabilization energies do not give any information about the possibility for a transfer from, e.g. a non-reconstructed diamond (100) surface to a 2 × 1-reconstructed (100) surface (i.e. the kinetics of a phase transformation). However, the activation energy (i.e. energy barrier) for this specific phase transformation was earlier studied by performing DFT calculations. A very small energy barrier of energy (0.32 eV) was obtained for the phase transformation from the

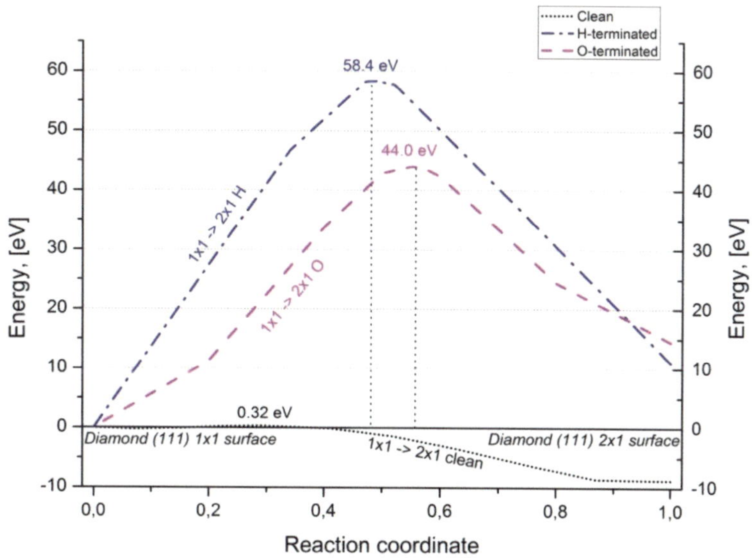

Figure 3.
Calculated activation energies for transformation between diamond (111), (1 × 1), and (2 × 1) structures as a function of surface termination with H and O, respectively.

non-reconstructed (111) surface to the slightly more stable 2 × 1-reconstruction of the same surface (also called a Pandey-Chain structure) (**Figure 3**). This minor barrier energy was recently justified by using ab initio molecular dynamic (MD) simulations on a non-reconstructed diamond (111) at 500 K [29].

4. Energetic stability of surface termination

4.1 General

The adsorption of various species onto diamond (111) and (100) surfaces were theoretically studied in earlier studies by the present author [28, 30]. The adsorption energies for the species O and OH are presented in **Figures 4** and **5**, where the adsorption and stabilization energies are shown over the whole range (0–100%) of surface coverage. Each of these figures starts with a 100% surface coverage with H adsorbates. These H adsorbates are then, one by one, replaced with oxygen atoms or hydroxyl molecules.

The stabilization energies in these figures are defined as the total adsorption for a 100% coverage of H, O, or OH, to be compared with the total adsorption energy for a 100% H-terminated diamond surface. A negative stabilization energy does, hence, mean that the corresponding surface configuration is energetically more favorable than a completely H-terminated surface.

4.2 Hydrogen termination

A clean surface was the starting point for the hydrogen adsorption studies. As was shown in Ref. 28, the clean (111)-2 × 1 surface is energetically more favorable than the corresponding 1 × 1 reconstruction (**Figure 6**). However, the (111)-1 × 1 phase becomes favorable for a hydrogen coverage above 30%. It has also experimentally been shown that a small number of hydrogen atoms on the diamond surface will reconstruct the (111)-2 × 1 to the 1 × 1 configuration [30]. A LEED pattern showed

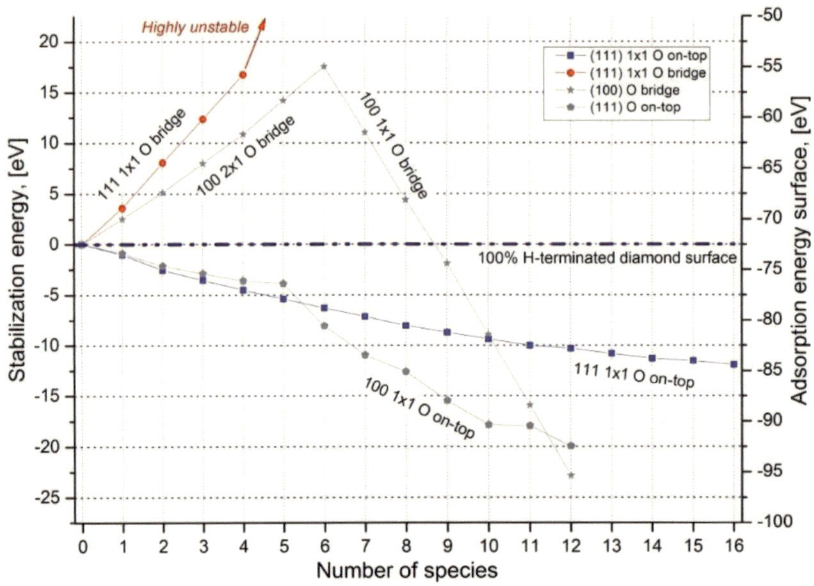

Figure 4.
An initially 100% H-terminated diamond (100)-2 × 1 or (111)-1 × 1 surface is successively oxygen-terminated by successively replacing the H adsorbates with O in either bridge or on-top positions.

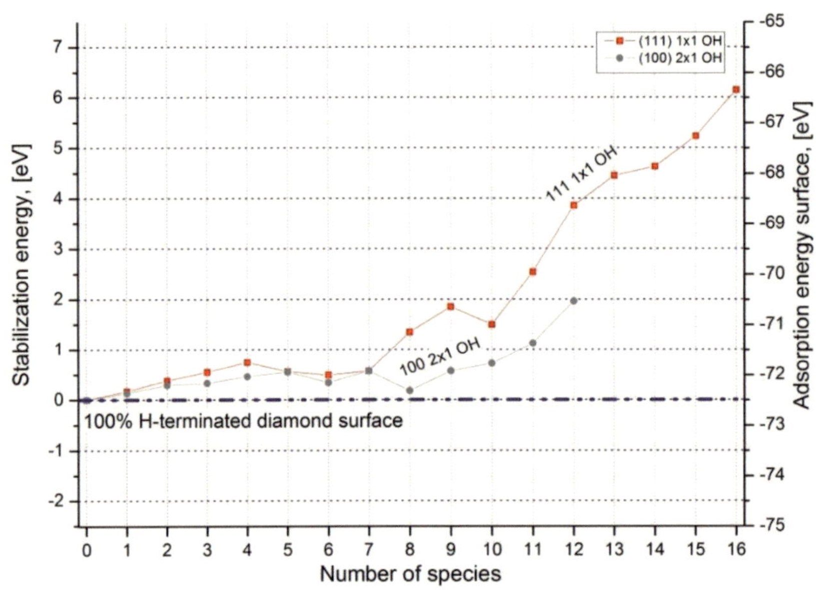

Figure 5.
Stabilization and adsorption energies for a successively replacement of H adsorbates with OH groups on diamond (100)-2 × 1 and (111)-1 × 1 surfaces.

that 0.05 ML would effectively reconstruct the (111)-2 × 1 surface to the 1 × 1 reconstruction. Hence, hydrogen atoms are not able to stabilize the Pandey chain carbon atoms as much as they do for the non-reconstructed 1 × 1 surface. A plausible explanation may be that the delocalized π-bonds of the Pandey chain are weakened. Hence, the electron configuration that stabilized the clean Pandey chain becomes perturbed.

The value of the adsorption energy for hydrogen on the diamond (100)-1 × 1 surface was in Ref. [30] found to be surprisingly large (−6.96 eV). Two hydrogens

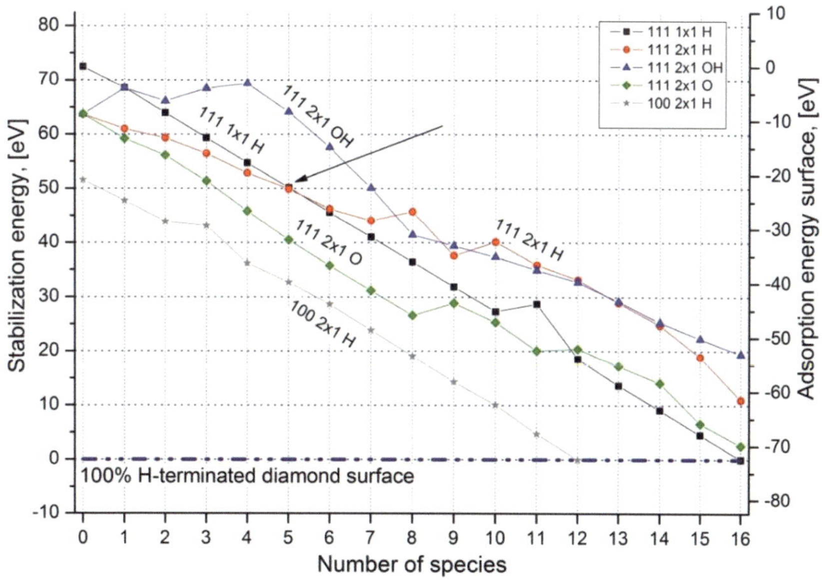

Figure 6.
A non-terminated 2 × 1 reconstructed diamond (111) surface successively H-, O-, and OH-terminated with increments of one adsorbate. A comparison with H-termination on diamond (111)-1 × 1 and (100)-2 × 1 is here included. The arrow identifies the H-coverage here; a (111)-1 × 1 and a (111)-2 × 1 surface have identical stabilization energy.

per carbon atom were chemisorbed onto the surface in order to saturate the two dangling bonds (per surface carbon) and thereby to uphold the sp^3 configuration and prevent 2 × 1 reconstruction. This dihydride configuration has been under much speculation [31–35], but it is generally accepted [36] that it does not occur during CVD growth conditions. The very favorable adsorption energy for the dihydride surface configuration is probably not seen experimentally due to the favorable (and spontaneous) 2 × 1 reconstruction. Moreover, the calculated adsorption energy for hydrogen on the 2 × 1-reconstructed (100) surface (−4.30 eV [30]) is in excellent agreement with experimental investigations [37].

In summary, the most stable H-terminated diamond surfaces phases under experimental conditions are pristine (111) and 2 × 1-reconstructed (100) (**Figure 6**).

4.3 Oxygen termination

Oxygen atoms are generally more effective than hydrogen in stabilizing a diamond surface. This is most probably due to the divalent nature of oxygen, whereby it may break C-C double bonds and form two C–O bonds. As a consequence, a high surface coverage of oxygen will yield a de-reconstruction of the 2 × 1-reconstruction of the diamond (100) surface into a bulk-equivalent 1 × 1 surface structure. This is shown in **Figure 4** for the termination with O atoms in both on-top and bridge positions [30]. As presented in Ref. 30, the most favorable position for oxygen on the (100) surface seems to be the bridge position which is supported by other experimental and theoretical studies [38–40]. The spontaneous reconstruction from (100)-2 × 1 to (100)-1 × 1 is shown in **Figure 4**. This implies that no activation barrier is present during this specific surface phase transformation process (since DFT is a 0 K method).

For the situation with a diamond (111) surface, no surface de-reconstruction was observed as a result of oxygen adsorption (100% coverage) onto the (111)-2 × 1 surface. Hence, the (111)-2 × 1 to (111)-1 × 1 reconstruction must be associated with an activation energy. This is true for also the reverse direction. Furthermore, the on-top and bridge positions of O were only found stable for an oxygen coverage less than 50% (for both diamond (111)-2 × 1 and 1 × 1).

In summary, the most stable O-termination of a diamond (100) surface seems to take place for oxygen in bridge position on the bulk-like (100)-1 × 1 surface. For the situation with a diamond (111) surface, the most stable O-termination is most probably for an on-top position at a surface coverage of 100%.

4.4 Hydroxyl-termination

For the terminating OH groups, there are two types of adsorbate-adsorbate interactions: hydrogen bonding (energy stabilization) and steric repulsions (energy destabilization). The hydrogen bonding will be noticeable at lower OH coverage, whereas steric repulsion dominates at higher coverage. The destabilization of OH-terminated surfaces is supported by an experimental study that reports a small number of hydroxyl groups present on an oxygenated (100) surface [1]. Hydrogen-bonding enthalpies of −0.15 eV were thereby reported.

The adsorption of OH groups onto diamond surfaces was calculated in Ref. 30, using a DFT method. The results for a 2 × 1-reconstructed (100) were found to be −4.1 eV. Hence, this adsorption energy is in good agreement with the corresponding results for the 1 × 1 surface: −4.2 eV (0.1 eV difference). Another DFT study for the diamond (111) surface has reported almost identical results for the OH adsorption energies of −4.2 (−4.0) eV for the 1 × 1 (and 2 × 1) reconstruction.

As shown in Ref. 28, the theoretically calculated adsorption energies for OH groups adsorbed onto the Pandey chain are unfavorable up to a 30% surface coverage (see **Figure 5**). A severe weakening of the π-bond chain, as well as absence of hydrogen bonding, is probably responsible for this effect.

4.5 Fluorine-termination

The adsorption of fluorine onto diamond (100) and (111) surfaces in their most common surface reconstructions (1 × 1 and 2 × 1) was also studied in Refs. 28 and 30. Only 100% terminated surfaces were considered. Fluorine is the termination species that most resembles hydrogen: monovalent with a rather small covalent radius. Fluorine was found to be strongly adsorbed to all of the studied diamond surfaces (−5.53, −6.33, −4.56, and −4.56 eV for the (100)-1 × 1, (100)-2 × 1, (111)-1 × 1, and (111)-2 × 1 surfaces, respectively). Although the diamond (100)-1 × 1 surface has two unpaired electrons, two fluorine atoms per carbon (difluoride configuration) atom was in Ref. 30 found to be highly unstable.

5. Effect by terminating species on the diamond surface electronic structure

The diamond surface electron structure and properties, such as electron affinity (EA) and has work function (WF), have earlier been studied both experimentally and theoretically in a vast number of publications. The diamond surfaces included both (111) and (100) surface planes, and the dominant surface-terminating species were hydrogen (H), oxygen, as well as fluorine. A mixture of O-containing species is most probably present on the surface, and it is not yet outlined how the atomic-level structure looks like for fluorinated surfaces. However, what is missing is a more detailed information about the influence of N-containing adsorbates on both the (111) and (100) diamond surfaces. This is also the situation for, theoretically, the extreme situation with a 100% coverage of O_{ontop}, O_{bridge}, OH, F or NH_2, species. The aim with a recent study by the present author was, hence, to theoretically investigate the effect by these individual terminating species on especially the EA of diamond (111) and the 2 × 1-reconstructed (100) surface. This theoretical study was based on DFT calculations, and the results are presented in **Table 1**.

At first, the DFT method used in calculating the electronic structures was evaluated by comparing the calculated band gap for bulk diamond with the experimental one: 5.4 vs. 5.5 eV. These band gaps were found to be almost identical, and, hence, the present DFT method was chosen for all of the density of state (DOS) calculations in the present study.

As a result of the geometry optimizations, it was not possible to bind O in bridge position on the diamond (111) surface. These O atoms were in the course of the optimization immediately transferred to on-top positions instead. Moreover, the (111) surface stayed in this state, and did not reconstruct to the Pandey Chain reconstruction when being adsorbed to. This was also the situation for the 2 × 1-reconstructed (100) surface and for all terminating situations except for O_{ontop}^{-} adsorption. This specific adsorption induced a re-reconstruction to diamond (100)-1 × 1. The reason for this process is that O forms a double bond with the surface carbon, thereby breaking the C-C bond within the paired C couple on the reconstructed (100) surface.

As can be seen in **Table 1**, the adsorption of H, OH, and NH_2 resulted in negative electron affinities (NEA) for both diamond (111) and (100); −1.0, −0.4, and − 2.0 eV for diamond (111); and −0.7, −1.3, and −0.9 eV for diamond (100). For comparison, the experimentally obtained value of NEA for H is about −1.1. Hence, the here presented calculation of EA for H adsorption shows a very good resemblance with experimental findings. Moreover, adsorption of F, O_{ontop}, and O_{bridge} resulted in positive electron affinities (PEA): +1.1 (F) eV for diamond (111) and +2.1, +3.8, and +0.4 eV for diamond (100). Also, O_{ontop} adsorption onto diamond (111) resulted in a metallic surface (i.e. surface electron conductivity). For comparison, the experimentally obtained value of PEA for oxygen is about +1.7. Hence, the here presented calculation of EAs for oxygen adsorption shows a not so good resemblance with experimental findings. But it must here be stressed that there is most probably a mixture of O-species on the surface, and it is not possible to analyze (experimentally) this distribution more in detail.

eV	H	OH	F	O_{ontop}	O_{bridge}	NH_2
(111)						
VBE	-4.0	-4.0	-5.8	Cond.	---	-2.3
EA	-1.0	-0.4	+1.1	Cond.	---	-2.0
(100)						
VBE	-4.8	-3.9	-7.0	-4.3	-4.4	-3.4
EA	-0.7	-1.3	+2.1	+3.8	+0.4	-0.9

The values are given in eV.

Table 1.
Positions of the valence band edge (VBE) and value of electron affinity (EA) for variously terminated diamond (111) and (100) surfaces.

6. Effect by doping on the diamond surface electronic structure

6.1 General

Diamond, as a material, possesses quite unique electrochemical properties, such as a large electrochemical potential window, a low dielectric constant, controllable surface termination, and a high breakdown voltage [41]. However, the large band gap (5.4 eV) limits the usage of diamond for electronic devices. Hence, impurities have been introduced into the diamond lattice, which can dramatically change both the geometrical and electronic properties. The boron atom is one of the most commonly used p-type dopants. It has been found that the B dopant will increase the electronic conductivity, as compared with the intrinsic non-doped diamond [1]. For different doping levels, the diamond material will show a metallic-type conductivity [42, 43], p-type semi-conductivity [44, 45], and superconductivity [46]. Nitrogen and phosphorus are commonly used as n-type dopants within the diamond material. However, substitutional doping with N atoms creates a deep donor level in diamond, which limits the usage of this material as an n-type semiconductor [3–6]. Instead, substitutional doping with phosphorus is regarded to be more efficient in creating an n-type semiconductor [7, 47].

6.2 Substitutional B doping

Substitutional doping with boron atoms did not render any major structural changes in the diamond surface region, as compared to the corresponding non-doped situations. Moreover, the atomic dopant was positioned within the second C layer, and this was the situation for all three different types of dopants studied (i.e. B, N, and P). The reason to this position is that these dopants have earlier been found to have a major influence on the surface structure and properties when positioned in C layer 2.

As can be seen in **Table 1**, the adsorption of H, OH, and NH_2 resulted in NEAs for B-doped diamond (111): −0.6, −0.3, and −0.6 eV. It was, however, only H-terminated diamond (100) surfaces that resulted in a NEA of −0.6. The other terminating species (OH and NH_2) resulted in PEAs of +0.1 and +3.2 eV. The results obtained for the (111) surface are thus similar to the non-doped situation, whilst it is completely different for the (100) surface.

As can further be seen in **Table 1**, the adsorption of F species on both the (111) and (100) B-doped diamond surfaces resulted in PEA values of +0.8 vs. +1.1 eV. Moreover, the influence by B doping was found to be very small. As stated above, adsorption of O in bridge position was only feasible for the diamond (100) surface. The calculated PEA value for the B-doped surfaces was +0.7 eV, which is also very similar to the corresponding value for a non-doped surface. On the contrary, the calculated results for O_{ontop} position were completely different when comparing diamond (111) with (100). Both non- and B-doped (111) surfaces rendered surface conductivity (i.e. a metallic-like situation), whilst the PEA values obtained for the (100) surfaces were completely different with respect to each other: +3.8 (non-doped) vs. +1.1 (B-doped) eV. An experimentally obtained EA value for the oxygen-terminated non-doped diamond surface is about +1.7 eV. However, it should be kept in mind that a surface coverage with oxygen most often involves a variety of O-related species (e.g. O and OH), as well as different types of adsorption sites (e.g. O_{ontop} and O_{bridge}). It is therefore very difficult to compare experimental values with theoretically obtained ones.

6.3 Substitutional N doping

As was the situation with B doping, substitutional doping with nitrogen atoms within the second diamond C layer did not render any major structural changes in the diamond surface region, as compared to the corresponding non-doped situations.

As can be seen in **Table 1**, the adsorption of H, OH, and NH_2 resulted in NEAs for N-doped diamond (111): −0.5, −0.1, and − 0.8 eV. On the contrary, these species resulted in an electrically conductive surface for diamond (100). Hence, the valence bands were not filled when using these terminating species in combination with substitutional N doping. Moreover, the other terminating species (O_{ontop} and F) on the diamond (111) species resulted also in an electrically conductive surface. The results obtained for the (111) surface are thus, with one exception, similar to the non- (or B-) doped situation. F-termination on a non- (or B-)doped diamond (111) surface resulted in a PEA, whilst it gave surface conductivity for the corresponding N-doped situation.

When comparing diamond (111) with (100) in **Table 1**, it is obvious that only O_{ontop}-termination yielded similar results for the two different surface planes (i.e. surface electronic conductivity). The other surface-terminating species gave completely different results. For O_{bridge}-termination on diamond (100), which cannot be found on diamond (111), it was shown that O_{bridge}-termination in combination with B doping will render a PEA value of +2.6 eV. This value is closer to the experimental value for non-doped diamond (100) (+1.6 eV) than to the calculated one (+0.4 eV). When comparing an N-doped diamond (100) surface with non- (or B-) doped counterparts, it has been found that only O_{bridge}- and F-termination will give similar results (i.e. PEA values). The other terminating species (H, OH, O_{ontop}, and NH_2) give completely different results.

6.4 Substitutional P doping

As was the situation with both B and N doping, substitutional doping with phosphorus atoms within the second diamond C layer did not render any major structural changes in the diamond surface region, as compared to the corresponding non-doped situations.

As can be seen in **Table 1**, the adsorption of all adsorbates (H, F, O_{ontop}, O_{bridge}, OH, and NH_2) resulted in electrically conductive surfaces for diamond (111). On the contrary, almost all of these species resulted in PEAs for diamond (100): +0.5 (H), +2.2 (F), +3.9 (O_{ontop}), +3.0 (O_{bridge}), and + 0.8 (OH) eV. The only adsorbate that did not render a PEA value was NH_2, which instead resulted in a NEA value (−0.7 eV). The results obtained for the (111) surface are thus, with only one exception, completely different compared to the non-, B-, or N-doped situation. O_{ontop}-termination gave surface electronic conductivities for both non-doped and doped diamond (111) surfaces. F-termination resulted in surface conductivity for both P- and N-doped diamond (111), whilst non- and B-doped surfaces showed PEA values. Moreover, H, OH, and NH_2 adsorbates induced NEA values, which is completely different to the surface conductivity observed for P-doped diamond (111).

Surface-terminated P-doped diamond (100) surfaces gave results that were both similar and dissimilar to the other types of doping situations. F and O_{bridge}-termination gave PEA values for all non-doped and doped situations (B, N, and P). O_{ontop}-termination resulted in PEA values for non-, B-, and P-doped surfaces, whilst it for N-doped diamond gave surface conductivity. As can be seen in **Table 1**, H-, OH-, and NH_2-termination resulted in a mixture of all three types of surface electronic properties (i.e. surface conductivity, PEA, and NEA), depending on the type of doping situation.

7. Conclusions

The combined effect of surface termination and substitutional doping on diamond surface electronic properties has here been presented and discussed. Firstly, the stability and geometrical structures of variously terminated diamond (111) and (100) surfaces were presented. The terminating species included H, F, OH, O_{ontop}, O_{bridge}, and NH_2. Moreover, the results for the doped surfaces were compared with non-doped diamond (111) and (100) surfaces. The doping species included B, N, and P. Theoretical DFT calculations had been used for the here presented results.

As a result of these first-principle calculations, F- and O_{bridge}-terminations gave PEA values for all non-doped and doped situations. Moreover, O_{ontop}-termination resulted in PEA values for non-, B-, and P-doped surfaces, whilst it for N-doped diamond gave surface conductivity. Various results were obtained for the other combinations of surface terminations and doping situations.

Acknowledgements

This work was supported by the Swedish Research Council (VR). The computational results were obtained using Materials Studio from BIOVIA.

Author details

Karin Larsson
Department of Chemistry-Angstrom Laboratory, Uppsala University, Uppsala, Sweden

*Address all correspondence to: karin.larsson@kemi.uu.se

References

[1] Shenai K, Scott RS, Baliga BJ. Optimum semiconductors for high-power electronics. Electron Devices IEEE Transactions. 1989;**36**(9): 1811-1823. DOI: 10.1109/16.34247

[2] Panizza M, Cerisola G. Application of diamond electrodes to electrochemical processes. Electrochimica Acta. 2005;**51**(2):191-199. DOI: 10.1016/j.electacta.2005.04.023

[3] Bhattacharyya S, Auciello O, Birrell J, Carlisle JA, Curtiss LA, et al. Synthesis and characterization of highly-conducting nitrogen-doped ultrananocrystalline diamond films. Applied Physics Letters. 2001;**79**(10):1441-1443. DOI: 10.1063/1.1400761

[4] Rohrer E, Graeff CF, Janssen R, Nebel CE, Stutzmann M, et al. Nitrogen-related dopant and defect states in CVD diamond. Physical Review B. 1996;**54**(11):7874-7880. DOI: 10.1103/PhysRevB.54.7874

[5] Yu BD, Miyamoto Y, Sugino O. Efficient n-type doping of diamond using surfactant-mediated epitaxial growth. Applied Physics Letters. 2000;**76**(8):976-978. DOI: 10.1063/1.125910

[6] Haase A, Peters A, Rosiwal S. Growth and thermoelectric properties of nitrogen-doped diamond/graphite. Diamond and Related Materials. 2016;**63**:222-226. DOI: 10.1016/j.diamond.2015.10.023

[7] Koizumi S, Kamo M, Sato Y, Ozaki H, Inuzuka T, et al. Growth and characterization of phosphorous doped {111} homoepitaxial diamond thin films. Applied Physics Letters. 1995;**71**: 1065-1067. DOI: 10.1063/1.119729

[8] Frangieh G, Jomard F, Pinault MA, Barjon J. Influence of tertiarybutylphosphine (TBP) addition on the CVD growth of diamond. Physica Status Solidi (a). 2009;**206**(9):1996-1999. DOI: 10.1002/pssa.200982225

[9] Kato H, Yamasaki S, Okushi H. Growth and characterization of phosphorus-doped diamond using organophosphorus gases. Physica Status Solidi (a). 2005;**202**(11):2122-2128. DOI: 10.1002/pssa.20561925

[10] Matsumoto T, Kato H, Makino T, Ogura M, Takeuchi D, et al. Carrier transport in homoepitaxial diamond films with heavy phosphorus doping. Japanese Journal of Applied Physics. 2014;**53**:5. DOI: 10.7567/JJAP.53.05FP05

[11] Grotjohn TA, Tran DT, Yaran MK, Demlow SN, Schuelke T, et al. Heavy phosphorus doping by epitaxial growth on the (111) diamond surface. Diamond and Related Materials. 2014;**44**:129-133. DOI: 10.1016/j.diamond.2014.02.009

[12] Hoshino Y, Kato H, Makino T, Ogura M, Iwasaki T, et al. Electrical properties of lateral p–n junction diodes fabricated by selective growth of n+ diamond. Physica Status Solidi (a). 2012;**209**(9):1761-1764. DOI: 10.1002/pssa.201200053

[13] Oyama K, Sakai N, Yamasaki S, Uchida K, Takeuchi D, et al. Enhancement in emission efficiency of diamond deep-ultraviolet light emitting diode. Applied Physics Letters. 2011;**99**(6):061110. DOI: 10.1063/1.3625943

[14] Takeuchi D, Makino T, Kato H, Ogura M, Okushi H, et al. High-voltage vacuum switch with a diamond p–i–n diode using negative electron affinity. Japanese Journal of Applied Physics. 2012;**51**(9R):090113. DOI: 10.1143/JJAP.51.090113

[15] Suzuki M, Yoshida H, Sakuma N, Ono T, Sakai T, et al. Electrical

characterization of phosphorus-doped n-type homoepitaxial diamond layers by Schottky barrier diodes. Applied Physics Letters. 2004;**84**:2349-2351. DOI: 10.1063/1.1695206

[16] Kato H, Takeuchi D, Ogura M, Yamada T, Kataoka M, et al. Heavily phosphorus-doped nano-crystalline diamond electrode for thermionic emission application. Diamond and Related Materials. 2015;**63**:165-168. DOI: 10.1016/j.diamond.2015.08.002

[17] Marcon L, Spriet C, Coffinier Y, Galopin E, Rosnoblet C, et al. Cell adhesion properties on chemically micropatterned boron-doped diamond surfaces. Langmuir: ACS Journal of Surfaces and Colloids. 2010;**26**(19):15065-15069. DOI: 10.1021/la101757f

[18] Yamada T, Yamaguchi H, Namba A, Okano K, Takeuchi D, et al. Field emission mechanism of oxidized highly phosphorus-doped homoepitaxial diamond (111). Applied Physics Letters. 2005;**87**(23):234107-234110. DOI: 10.1063/1.2135875

[19] Yamada T, Masuzawa T, Mimura H. Electron emission from conduction band of heavily phosphorus doped diamond negative electron affinity surface. Journal of Physics D. 2015;**49**(4):045102-045106. DOI: 10.1088/0022-3727/49/4/045102

[20] De Theije F, Roy O, van der Laag NJ, van Enckevort WJP, et al. Oxidative etching of diamond. Diamond and Related Materials. 2000;**9**(3):929-934. DOI: 10.1016/S0925-9635(99)00239-3

[21] Mei W, Simon N, Decorse-Pascanut C, Bouttemy M, Etcheberry A, et al. Comparison of the chemical composition of boron-doped diamond surfaces upon different oxidation processes. Electrochimica Acta. 2009;**54**:5818-5824. DOI: 10.1016/j.electacta.2009.05.037

[22] Ferro S, De Battisti A. The 5-V window of polarizability of fluorinated diamond electrodes in aqueous solutions. Analytical Chemistry. 2003;**75**(24):7040-7042. DOI: 10.1021/ac034717

[23] Kondo T, Ito H, Kusakabe K, Ohkawa K, Yasuaki E, et al. Plasma etching treatment for surface modification of boron-doped diamond electrodes. Electrochimica Acta. 2007;**52**(11):3841-3848. DOI: 10.1016/j.electacta.2006.11.001

[24] Song K-S, Degawa M, Nakamura Y, Kanazawa H, Umezawa H, et al. Surface-modified diamond field-effect transistors for enzyme-immobilized biosensors. Japanese Journal of Applied Physics. 2004;**43**(6B):L814-L817. DOI: 10.1143/JJAP.43.L814

[25] Vanderbilt D. Soft self-consistent pseudopotentials in a generalized eigenvalue formalism. Physical Review B. 1990;**41**(11):7892-7895. DOI: 10.1103/PhysRevB.41.7892

[26] Perdew JP, Burke K, Ernzerhof M. Generalized gradient approximation made simple. Physical Review Letters. 1996;**77**(18):3865-3868. DOI: 10.1103/PhysRevLett.77.3865

[27] Petrini D, Larsson K. Origin of the reactivity on the nonterminated (100), (110), and (111) diamond surfaces: An electronic structure DFT study. Journal of Physical Chemistry C. 2008;**112**(37):14367-14376. DOI: 10.1021/jp711190r

[28] Petrini D, Larsson K. Theoretical study of the thermodynamic and kinetic aspects of terminated (111). Journal of Physical Chemistry C. 2008;**112**(8):3018-3026. DOI: 10.1021/jp709625a

[29] Song Y, Larsson K. Formation Conditions for Epitaxial Graphene on Diamond (111) Surfaces. Accepted for

Publication in Advances in Natural Sciences. 2019

[30] Petrini D, Larsson K. A theoretical study of the energetic stability and geometry of hydrogen- and oxygen-terminated diamond (100) surfaces. Journal of Physical Chemistry C. 2007;**111**(2):795-801. DOI: 10.1021/jp063383h

[31] Yamada T, Chuang TJ, Seki H, Mitsuda Y. Chemisorption of fluorine, hydrogen and hydrocarbon species on the diamond C(111) surface. Molecular Physics. 1992;**76**(4):887-908. DOI: 10.1080/00268979200101741

[32] Yang SH, Drabold DA, Adams JB. *Ab initio* study of diamond C(100) surfaces. Physical Review B. 1993;**48**(8):5261-5264. DOI: 10.1103/PhysRevB.48.5261

[33] Yang YL, D'Evelyn MP. 38th National Symposium of the American Vacuum Society (AVS); 11th–15th of November 1991; Seattle, Washington, USA

[34] Zheng XM, Smith PV. The topologies of the clean and hydrogen-terminated C(100) surfaces. Surface Science. 1991;**256**(1-2):1-8. DOI: 10.1016/0039-6028(91)91194-3

[35] Kubiak GD, Hamza AV, Stulen RH, Sowa EG, Kolasinski KW. Hydrogen desorption and subsequent reconstruction on natural diamond surfaces. Carbon. 1990;**28**(6):751-752. DOI: 10.1016/0008-6223(90)90268-4

[36] Thoms BD, Butler JE. HREELS and LEED of H/C(100): The 2 × 1 monohydride dimer row reconstruction. Surface Science. 1995;**328**(3):291-301. DOI: 10.1016/00396028(95)00039-9

[37] Butler JE, Woodin RL, Brown LM, Fallon P. Thin film diamond growth mechanisms. Philosophical Transactions of the Royal Society A. 1993;**342**(1664):209-224. DOI: 10.1098/rsta.1993.0015

[38] Pehrsson PE, Mercer TW. Oxidation of the hydrogenated diamond (100) surface. Surface Science. 2000;**460**(1-3):49-66. DOI: 10.1016/S0039-6028(00)00494-5

[39] Frenklach M, Huang D, Thomas RE, Rudder RA, Markunas RJ. Activation energy and mechanism of CO desorption from (100) diamond surface. Applied Physics Letters. 1993;**63**(22):3090. DOI: 10.1063/1.110217

[40] Badziag P, Verwoerd WS. MNDO analysis of the oxidised diamond (100) surface. Surface Science. 1987;**183**(3):469-483. DOI: 10.1016/S0039-6028(87)80222-4

[41] Li HD, Li H, Li L, Zhang T, Lü X, et al. Investigation on crystalline structure, boron distribution, and residual stresses in freestanding boron-doped CVD diamond films. Journal of Crystal Growth. 2010;**312**(12-13): 1986-1991. DOI: 10.1016/j.jcrysgro.2010.03.020

[42] McCreery RL. Advanced carbon electrode materials for molecularelectrochemistry. Chemical Reviews. 2008;**108**(7):2646-2687. DOI: 10.1021/cr068076m

[43] Pleskov YV, Yu V, Sakharova AY, Krotova MD, Bouilov LL, et al. Photoelectrochemical properties of semiconductor diamond. Journal of Electroanalytical Chemistry. 1987;**228**(1-2):19-27. DOI: 10.1016/0022-0728(87)80093-1

[44] Swain GM, Ramesham R. The electrochemical activity of boron-doped polycrystalline diamond thin-film electrodes. Analytical Chemistry. 1993;**65**(4):345-351. DOI: 10.1021/ac00052a007

[45] Williams AW, Lightowl E, Collins AT. Impurity conduction in synthetic semiconducting diamond.

Journal of Physics C: Solid State PhysicsJournal of Physics C: Solid State Physics. 1970;**3**(8):1727-1735. DOI: 10.1088/0022-3719/3/8/011

[46] Chrenko RM. Boron, Dominant Acceptor in Semiconducting Diamond. Physical Review B. 1973;7(10): 4560-4567. DOI: 10.1103/ PhysRevB.7.4560

[47] Kato H, Futako W, Yamasaki S, Okushi H. Growth of phosphorus-doped diamond using tertiarybutylphosphine and trimethylphosphine as dopant gases. Diamond and Related Materials. 2005;**14**(3):340-343. DOI: 10.1016/j. diamond.2004.11.032